Jie Zeng, Xin Su, Bin Ren and Lin Liang (Eds.)
Multiple Access Technologies for 5G

Also of Interest

5G
An Introduction to the 5th Generation Mobile Networks
Ulrich Trick, 2021
ISBN 978-3-11-072437-0, e-ISBN (PDF) 978-3-11-072450-9,
e-ISBN (EPUB) 978-3-11-072462-2

Circularly Polarized Antenna Technology
Yufeng Wang, 2020
Together with: National Defense Industry Press
ISBN 978-3-11-056118-0, e-ISBN (PDF) 978-3-11-056280-4,
e-ISBN (EPUB) 978-3-11-056156-2

Communication Electronic Circuits
Zhiqun Cheng, Guohua Liu, 2020
Series: Information and Computer Engineering
Together with: China Science Publishing & Media Ltd.
ISBN 978-3-11-059538-3, e-ISBN (PDF) 978-3-11-059382-2,
e-ISBN (EPUB) 978-3-11-059293-1

De Gruyter Series on the Applications of Mathematics in
Engineering and Information Sciences
Edited by: Mangey Ram
ISSN 2626-5427, e-ISSN 2626-5435

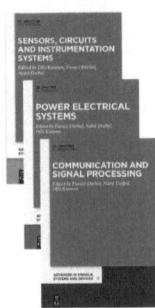

Advances in Systems, Signals and Devices
Edited by: Olfa Kanoun
ISSN 2364-7493, e-ISSN 2364-7507

Multiple Access Technologies for 5G

New Approaches and Insight

Edited by
Jie Zeng, Xin Su, Bin Ren and Lin Liang

人民邮电出版社
POSTS & TELECOM PRESS

Editors

Dr. Jie Zeng
Department of Electronic Engineering
Tsinghua University
100084 BEIJING
China
zengjie@mail.tsinghua.edu.cn

Dr. Bin Ren
DATANG Mobile
29 College Road
100083 BEIJING
China
renbin@datangmobile.cn

Prof. Xin Su
Research Institute of Information Technology
Tsinghua University
100084 BEIJING
China
suxin@wireless.mdc.tsinghua.edu.cn

Lin Liang
China Telecom
China Telecom Beijing Information Technology
Innovation Park
102209 BEIJING
China
lianglin@ctbri.com.cn

ISBN 978-3-11-066581-9
e-ISBN (PDF) 978-3-11-066636-6
e-ISBN (EPUB) 978-3-11-066597-0

Library of Congress Control Number: 2021934543

Bibliographic information published by the Deutsche Nationalbibliothek
The Deutsche Nationalbibliothek lists this publication in the Deutsche Nationalbibliografie;
detailed bibliographic data are available on the Internet at http://dnb.dnb.de.

© 2021 Posts and Telecom Press and Walter de Gruyter GmbH, Berlin/Boston
Cover image: Lisa-Blue/E+/Getty Images
Typesetting: Integra Software Services Pvt. Ltd.
Printing and binding: CPI books GmbH, Leck

www.degruyter.com

Contents

Chapter 1
Outline of Multiple Access Technology

1.1 History of Multiple Access Technology

Jie Zeng

Multiple access is a technique to implement multi-user communications by letting multiple users share one common channel, and it is one of the core physical layer techniques in wireless communications. The orthogonal resource allocation method is always applied in multiple access techniques, such as the Frequency Division Multiple Access (FDMA) in the first generation (1G) communication systems, the Time Division Multiple Access (TDMA) in the second generation (2G) communication systems, the Code Division Multiple Access (CDMA) in the third generation (3G) communication systems and the Orthogonal Frequency Division Multiple Access (OFDMA) in the fourth generation (4G) communication systems. However, when facing 2020 and the future, besides improving the system frequency efficiency, the fifth generation (5G) communication systems also need to support massive user equipment connections. This means more intelligent and higher efficiency resource allocation techniques need to be considered and the new multiple access (NMA) technique is thus proposed.

1.1.1 Frequency Division Multiple Access

In the 1980s, 1G analog cellular mobile communication network achieved commercial-scale application. Typical systems in 1G include the Advanced Mobile Phone System (AMPS) in the USA, the Total Access Communications System (TACS) in England and the Nordic Mobile Telephony System (NMTS) in northern Europe. The first generation (1G) uses analog modulation and FDMA as the main techniques. The principle of FDMA is illustrated in Figure 1.1. By cutting the total frequency channel into several uncorrelated sub-channels and allocating different sub-channels to different users, different users can be allocated in the same slot with different frequencies. Protection frequency band usually needs to be configured between sub-channels to resist nonideal filter, adjacent channel interference and frequency spread caused by Doppler shift [1]. Due to the drawbacks of FDMA-based analog cellular systems, such as low spectral efficiency, security risks and weak interference resistance ability, it was abandoned as a result of the market competition.

https://doi.org/10.1515/9783110666366-001

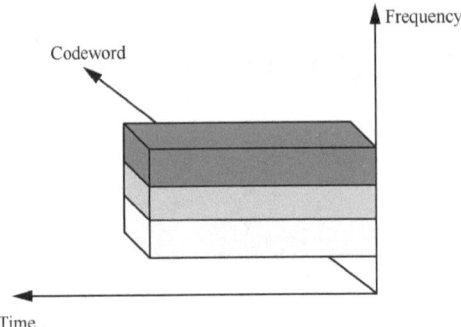

Figure 1.1: FDMA technique.

1.1.2 Time Division Multiple Access

The 2G system, a narrowband digital cellular system and represented by the Digital AMPS (DAMPS) in the USA and Global System for Mobile communications (GSM) in Europe, is widely used in the 1990s, which applies TDMA and narrowband CDMA. As illustrated in Figure 1.2, the transmission time is divided into different slots in TDMA, the transmission/reception of the user occupies one respective slot. However, different users cannot transmit simultaneously, in other words, different users transmit in orthogonal slots.

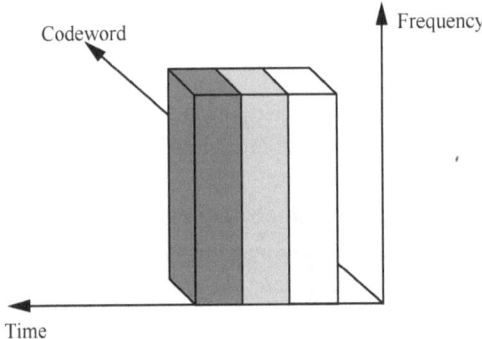

Figure 1.2: TDMA technique.

The main challenge for TDMA is the synchronization requirement among different users. In the downlink, all the signals are generated by the same transmitter and arrived at the receiver after passing through the same channel. For flat fading channels, as long as the users transmit in orthogonal time slots, the signals at the receiver maintain this orthogonality in time. However, in the uplink, signals from different users experience different channels with different delays, then TDMA in the uplinks needs synchronization to ensure the receiving signals are orthogonal in time. Synchronization usually demands cooperation between the base station (BS) and the access point (AP), which may cause large overhead. As a result, TDMA introduces a guard band between channels to compensate influences caused by multipath and synchronization errors [1].

1.1.3 Code Division Multiple Access

In 1996, International Telecommunication Union (ITU) named the standard of 3G as IMT-2000. The main types of 3G include CDMA2000, WCDMA and TD-SCDMA, of which CDMA is the base technique for 3G. In CDMA, pseudo-random sequence with bandwidth which is much larger than the signal bandwidth is used to spread the original signal, which is sent after carrier modulation. At the receiver, to recover the original signal, same pseudo-random sequence is used to de-spread the wideband signal. As illustrated in Figure 1.3, different users choose different spreading codes and occupy same time-frequency resources. Although one user can receive the superposed signals from all the users, the signals of the other users cannot be decoded.

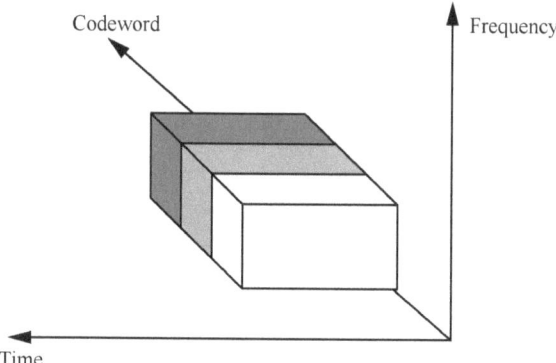

Figure 1.3: CDMA technique.

CDMA applies orthogonal spreading code in the downlink, such as Walsh–Hadamard code, and non-orthogonal spreading code in the uplink. By using non-orthogonal spreading code, the number of access users can be increased in the uplink, meanwhile the inter-user interference is introduced. Furthermore, the levels of the power at receiver are not the same due to the channel gains in the uplink are different for different users, which means that power control is needed to avoid the near–far effect [1].

1.1.4 Orthogonal Frequency Division Multiplexing Access

OFDMA is a multiple access method based on the technique of Orthogonal Frequency Division Multiplexing (OFDM). Since the subcarriers in OFDM are relatively independent of others, modulation method and transmit power can be assigned to each subcarrier specifically. As illustrated in Figure 1.4, OFDMA achieves multi-user access by allocating different subcarriers to a single user in different slots and distinguishing users through the subcarrier frequencies. On top of Figure 1.4, it can be seen that OFDMA is a combination of FDMA and TDMA. However, different from FDMA, OFDMA does not need to set guard bands between the frequency of different users, and the subcarriers of each user do not have to be continuous, which can greatly improve the frequency efficiency.

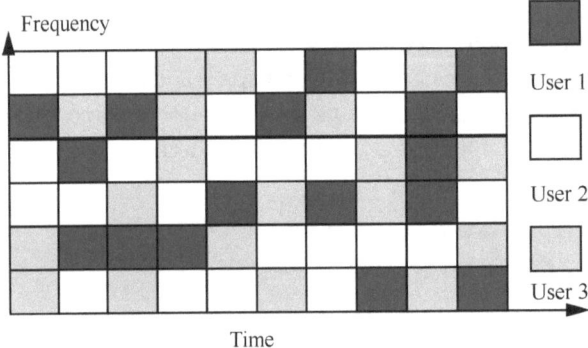

Figure 1.4: OFDMA technique.

1.1.5 New Multiple Access Technique in 5G

To satisfy the 5G network requirements, such as higher spectrum efficiency, larger capacity, more connectivity and lower latency, the resource utilization of 5G multiple access technique needs to be more flexible and efficient. The multiple access technologies from 1G to 4G mainly apply the orthogonal resource allocation method, whose access users are limited by resources and cannot satisfy the demands of massive access users in 5G. To support more users with limited resources, at the transmitter, 5G NMA technique implements non-orthogonal multi-user signal superposition in time-frequency resources by different domain procession (such as code domain, power domain, interleaving domain and so on), while at the receiver, multi-user signal separation by advanced multi-user detection technique is used. With user multiplexing in the same resource, NMA technique increases the number of access users in the network greatly. Thus, the total network capacity and spectrum efficiency are improved due to more access probabilities. Besides, NMA technologies can realize grant-free scheduling in a better way, which decreases the communication latency and user equipment (UE) power consumption. Compared to traditional orthogonal access techniques, NMA techniques can better fulfill the massive connection, low latency and low power consumption services in 5G.

Currently, the proposed candidate NMA techniques mainly include: Non-Orthogonal Multiple Access (NOMA), Pattern Division Multiple Access (PDMA), Sparse Code Multiple Access (SCMA), Multi-User Shared Access (MUSA) and Interleave-Grid Multiple Access (IGMA).

1.2 Current Research Progress of NMA Technology

Jie Zeng

1.2.1 Latest Progress on Standardization

As the specific organization for international radio communication standard formulation in ITU, ITU-Radio Communications Sector (ITU-R) has formulated the IMT-2000 (3G) standard and the IMT-Advanced (4G) standard successively. Now, ITU-R has started working on the 5G standard, and the Working Park 5D (WP5D) group is responsible for related issues. WP5D determined the work plan and time schedule for the research of IMT-2020. The detailed arrangements for the tasks and expected results of each meeting were also made. According to the plan, the requirement vision and technique trend of IMT-2020 should be accomplished in 2014; the time for collection preparation of 5G technique was carried out between 2015 and 2017; the collection was finished in 2018; the evaluation of 5G candidate techniques and the formulation of 5G standard has been done between 2019 and 2020. ITU-R introduced the technology trends of terrestrial IMT systems between 2015 and 2020 and beyond in the report "Future Technology Trends of Terrestrial IMT Systems" that was published in November 2014. The report pointed out that techniques such as advanced multiple access techniques could be applied to enhance the air interface performance. Besides, two typical NMA techniques including PDMA and SCMA are highlighted. In addition, WP5D work group mentioned the NMA techniques again in enhanced air interface techniques in "IMT Vision – Framework and Overall Objectives of the Future Development of IMT for 2020 and Beyond," which was published on July 21, 2015. It determined the work plan as follows: Detailed research on performance requirements, benchmark and methods for IMT new radio evaluation would be developed in 2016; related discussion meeting would be convened no later than the end of 2017, which would be responsible for the performance requirements, benchmark and methods discussion of the IMT-2020 candidate techniques, and provided an opportunity to discuss candidate techniques in informal situations for potential IMT-2020 supporters.

The 3rd Generation Partnership Project (3GPP) is an authoritative mobile communication specification determination organization. The Radio Access Network (RAN) work group, which is a subordinate of 3GPP, has already launched the discussion on NMA technique in the 3GPP RAN WG1 #84bis meeting convened at Pusan, South Korea in April 2016. At least six candidate NMA schemes are discussed in the meeting. According to incomplete statistics, many companies including Huawei, ZTE, CATT, NTT DoCoMo, Qualcomm and Alcatel-Lucent Shanghai Bell have submitted more than 20 proposals in the meeting. In the RAN WG1 #85 meeting held at Nanjing in May 2016, companies including Huawei, ZTE and CATT supplemented link-level simulation results to corresponding techniques, and companies including Samsung, LG and Fujitsu proposed their own NMA schemes. The total number of candidate NMA schemes was thus increased to 12. Besides, some other companies, such as Xinwei and Honxin, also took an active part in the research of NMA. The companies had reached two agreements on NMA [2]: NMA techniques should be studied aiming at different application scenarios and practical cases in new radio (NR); the NMA technique based on spontaneous grant-free scheduling should be studied at least in the scenario of uplink massive Machine Type Communications (mMTC). In the RAN WG1 #86 meeting held at Goteborg, Sweden, in August 2016, companies including Huawei, CATT and Samsung submitted the latest link-level and system-level simulation results on NMA schemes, and candidate schemes had increased to 15. Two suggestions were passed formally in this meeting [3]: NR needs to support non-orthogonal transmission in the uplink in addition to orthogonal schemes at least in mMTC; spontaneous grant-free scheduling is used in the uplink at least in mMTC. Furthermore, the research emphasis in the next step and system-level simulation parameters for NMA techniques were determined in this meeting.

1.2.2 Current Research Status of Different Countries

NMA technologies draw wide attention in China, and a lot of companies have proposed their own schemes. Standardization organizations such as IMT-2020 (5G) Promotion Group, China Communications Standards Association (CCSA) and the FuTURE Forum also paid attention to NMA technologies. IMT-2020 (5G) Promotion Group is founded by Ministry of Industry and Information Technology, Ministry of Science and Technology and National Development and Reform Commission. It is an organization that research the demands, frequencies, techniques and standards of 5G with industry, and the members of IMT-2020 (5G) Promotion Group come from telecom equipment manufacturers, telecom operators, universities and research institutes in China. As the main platform for domestic 5G promotion, IMT-2020 (5G) Promotion Group established a task group on non-orthogonal transmission in Beijing on December 20, 2013, and held the first meeting. Many NMA schemes were discussed in this meeting. The research directions, expected output reports and the preliminary work plan were

determined. In the second special topic conference opened on June 10, 2014, companies such as Huawei, ZTE and CATT attended and submitted the first draft of research
reports on NMA techniques. These reports summarized the application scenarios,
technical classifications, key technologies and technical evaluations of NMA initially.
As the research progresses, Huawei, ZTE and CATT submitted respective research
progress on NMA in the third and fourth meetings. In the fifth special topic meeting
held on December 8, 2014, non-orthogonal transmission task group was renamed to
NMA task group officially. The configuration of NMA technique simulation parameters was discussed during the meeting and uplink link-level simulation parameters
were determined. In the white paper on 5G concept published in February, 2015, IMT-
2020 (5G) Promotion Group pointed out that the NMA technique was one of the important innovation techniques in 5G wireless areas, and gave the specific definition of
the NMA technique. The white paper also introduced some NMA schemes. With more
and more companies and universities participating in the discussion and proposing
new schemes, the number of participants in the meeting and the received proposals
gradually increased. In recent meetings, more than 10 proposals from companies and
universities have been received. In the ninth and the tenth meetings, bit interleaving
code multiple access proposed by Tsinghua University, Non-Orthogonal Code Access
(NOCA) jointly proposed by Nokia and Shanghai Bell, and IGMA from Samsung have
formally joined the discussion of the NMA task group. Besides, the task group also
kept up with the pace of 3GPP, promoted the link-level simulation and system-level
simulation based on the 3GPP research progress and supported the 3GPP standardization and 5G technique research and develop (R&D) test.

In addition to the IMT-2020 (5G) Promotion Group, the cutting-edge wireless
technique work group (WG6) of CCSA wireless communication technical committee
(TC5) had a discussion on NMA technologies in the 38th meeting held in Shanghai
in September 2014. DoCoMo, Huawei, CATT and ZTE introduced the work principles,
preliminary performance evaluation results and advantages compared with orthogonal multiple access technique of their proposed new multiple technologies separately. The FuTURE Forum made an agreement on the importance of NMA techniques
in the meeting held in July 2014 and pointed out that NMA was a new assistance for
5G [4]. In the white paper published in 2015, the FuTURE Forum introduced flexible
access technique (the same as NMA technique) and gave a unified framework for
downlink and uplink [5]. Moreover, the FuTURE Forum released the white paper for
NMA technique. The white paper gives a detailed introduction on technique principle,
performance evaluation and system design of NMA technologies including Bit Division Multiplexing (BDM), MUSA, NOMA, PDMA, Resource Spread Multiple Access
(RSMA) and SCMA and summarizes the characteristics of different technologies comprehensively and authoritatively, and therefore has an important reference value for
the research of NMA.

The 2020 and Beyond Ad hoc (20B AH) in Japan is an organization established
by Association of Radio Industries and Businesses (ARIB) in August 2013 for the

research of terrestrial communication systems in 2020 and beyond. It consists of 31 ARIB members and is mainly responsible for the research on system concepts, basic functions, function distribution and architecture of future mobile communication systems. There are two subordinate work groups in 20B AH including the work group on service and concepts (WG-SC) and the work group on system architecture and wireless access technique (WG-Tech). The WG-Tech work group is mainly responsible for trending research of wireless access techniques and other key network techniques. The 20B AH gave a detailed introduction on the visions for IMT-2020 and future communication systems in "Visions for IMT 2020 and beyond" published on February 12, 2014, and indicated definitely that NMA is one of the key candidate techniques for 5G wireless access. In the "2020 and future mobile communication systems" white paper published on October 8, 2014, 20B AH mentioned that future wireless access systems meet requirements of increasing data services and massive accessing users. NMA technologies can realize data transmission from different groups on the same resource by the combination of power and spreading code. The candidate NMA techniques include NOMA, SCMA, Interleave Division Multiple Access (IDMA) and so on.

The 5G Forum in South Korea was founded in May 2013 with the vision: To become the global leader in 5G mobile communications before 2020. In the "5G Vision, Demands and Technologies" white paper published in February 2015, 5G Forum indicated that the NMA is one of the key techniques enabling the spectrum efficiency 100–1,000 times higher than before. The white paper also analyzed the two-user capacity region of the NMA, and proved that it can increase the system capacity when compared with orthogonal multiple access.

In America, 4G Americas gave some introductions on 5G requirements and candidate technologies in "4G America's Recommendations on 5G Requirements and Solutions" white paper published in October 2014 and "5G technique development suggestions" white paper published in October 2015. In these white papers, it is indicated that for multiple access technologies, orthogonal can avoid inter-user interference and ensure high system capacity. However, allocating orthogonal resources to different users may require a large number of signaling and bring extra delay to fast-access and small-payload services. Therefore, NMA can be used as a supplement to orthogonal multiple access in 5G systems. In the "Mobile Broadband Transformation: From LTE to 5G" white paper published in August 2016, 5G Americas indicated clearly that in 5G systems, NMA could be a supplement to orthogonal access with the help of advanced interference cancellation technologies.

In Europe, implementing "Framework Program" is the operation mode for 3G and 4G standard formulation. The Seventh Framework Program (FP7) operated between 2007 and 2013 had already promoted the 5G core topic research positively. The Eighth Framework Program (FP8) which started later changed its name to Horizon 2020 is the biggest framework program in European Union (EU), and funds continuously for 5G future research and innovation. The Mobile and Wireless

Communications Enablers for the Twenty-Twenty (2020) Information Society (METIS) program, which started officially at the end of 2012, is a 5G research program under the FP7 framework. Its goal is to build the foundations for 5G mobile and wireless communication systems, achieve agreements on requirements, characteristics and targets for future mobile communication and wireless technologies and form a unified opinion on concepts, embryos and key technique composition [6]. In the METIS program, Work Package (WP) is used to carry out the research and develop work on related domains and techniques. Each WP makes specific research aiming at different missions. WP2 is mainly responsible for developing and researching new wireless links to satisfy the demands of future application scenarios and user cases, building an agreement for the whole system design, and supporting a variety of large-scale services effectively. The NMA technique is one of its important missions. In "On new wireless access solutions" published on February 28, 2015, METIS mentioned that multiple access is one of the three important areas in METIS wireless link research. This article also indicated that non-orthogonal or semi-orthogonal multiple access technique is one of the main candidate techniques in future wireless communication systems. It also gave an introduction to NOMA and SCMA. Reference [7] presents the research of METIS on 5G mobile wireless system definition, and mentions that new techniques like NOMA can be used to satisfy the requirements of future 5G systems.

The 5G infrastructure Public–Private Partnership (5G PPP) is another 5G research program in Europe. Different from the basic technique research of METIS, 5G PPP pays more attention to the evolution of system standardization and industrialization. Meanwhile, the achievements of METIS will be imported to 5G PPP. In the "The Challenge, Research Emphasis and Suggestions of 5G" joint white paper published in September 2014, 5G PPP pointed out that radio resource allocation schemes based on non-orthogonal access can further lower the system synchronization requirement, and it needs to be further researched [8].

1.2.3 Current Research Status of Industry

As one of the earliest companies participating in 5G research, Huawei set up a joint research subject on 5G technology with foreign universities (such as Harvard University, University of California, Berkeley and University of Cambridge) since 2009, and took part in the EU METIS program as one of the sponsors. In the IMT-2020 NMA special topic conference held on June 9, 2014, Huawei gave a formal introduction about SCMA and showed some preliminary research results including the advantage, technical principle, grant-free access and link-level performance of SCMA. In the third special topic conference, Huawei submitted further research results about the SCMA blind detection. In the NMA special topic conference held on October 17, 2014, Huawei submitted the latest research results on SCMA codebook design, receiver scheme and link-level simulations. According to these results, in large-payload conditions, outer iteration of SCMA receiver can improve the receiving performance greatly with the performance bound almost coincides with single-user situation when the overload gain is 300%. In the following special topic conferences, Huawei updated research progress on SCMA and submitted uplink grant-free scheduling scheme, downlink access scheme, link-level simulation results and system-level simulation results. Besides, Huawei took part in the NMA discussion organized by 3GPP RAN WG1 positively. In the 3GPP RAN WG1 #84bis meeting held at Busan, South Korea, in April 2016, Huawei submitted proposals on technical principles, codebook design, codebook mapping, receiver schemes and performance advantages of SCMA. In RAN WG1 #85 and #86 meetings, Huawei attended and submitted proposals on the latest SCMA evaluation results and problems about grant-free scheme and user collision in NMA. Meanwhile, Huawei promoted the research of METIS and FuTURE Forum on NMA actively.

As one of the earliest companies that proposed NMA schemes, CATT is a main promoter of NMA. In the NMA special topic meeting opened on June 9, 2014, CATT introduced the work progress and work plan for Successive interference cancellation Amenable Multiple Access (SAMA). CATT had completed preliminary SAMA link-level verification in Single Input Multiple Output (SIMO) and Multiple Input Multiple Output (MIMO) situations with open-loop TM1 and TM3. Simulation results showed SAMA could achieve significant performance gain. In the third and the fourth NMA special topic conferences, CATT submitted research results on advanced receivers. In the CCSA TC5 WG6's 38th meeting held at Shanghai in September 2014, CATT proposed PDMA based on SAMA, in which users are distinguished by using non-orthogonal characteristic pattern at the transmitter, the semi-optimization multi-user detection is achieved by using Successive Interference Cancellation (SIC) at the receiver. In the fifth IMT-2020 NMA special topic conference held on December 18, 2014, CATT submitted preliminary link-level simulation results on PDMA, which indicated that PDMA can highly improve the Bit Error Rate (BER) performance compared with orthogonal transmission. In the next IMT-2020 NMA special topic conferences, CATT submitted the latest research progress, including PDMA system-level simulation results and develop and test plan of the verification platform. CATT also took part in the discussion on NMA organized by 3GPP RAN WG1 actively and submitted proposals on PDMA technical principle, the latest evaluation results and NMA evaluation methods. Furthermore, CATT took the lead in undertaking the 5G new modulation and coding and high-efficiency link technique research and development project in National High technology Research and Development Program of China (National "863" Project), in which the most important task is the research and application of PDMA key technology. Besides participating in the technology research and standard promotion of NMA, CATT also took part in the research discussion on NMA held by organizations such as the FuTURE Forum.

ZTE is also one of the earliest companies that joined the NMA research, which proposed and introduced MUSA in the fourth IMT-2020 NMA special topic conference.

Later on, ZTE participated in the research and discussion sponsored by standardization organizations including 3GPP, IMT-2020 and the FuTURE Forum. They updated and submitted MUSA simulation results continuously. In addition, ZTE proposed their own point of view on some aspects of NMA, such as spontaneous grant-free competitive access scheme [9], blind detection algorithms [10] and classification of NMA technologies [11].

Samsung participated in the research and standardization work on NMA sponsored by international organizations positively. In the 3GPP RAN WG1 #85 meeting, Samsung proposed IGMA scheme for the first time, and submitted link-level simulation results. In the 3GPP RAN WG1 #86 meeting afterward, Samsung submitted the latest evaluation results of IGMA and gave some research suggestions on grant-free collision, link adaptive and so on. Besides, Samsung took part in the NMA special topic conferences organized by China IMT-2020 Promotion Group actively.

In the RAN WG1 #84bis meeting, LG Electronics proposed Non-Orthogonal Coded Multiple Access (NCMA) scheme. In addition, in the white paper published on June 21, 2016, LG Electronics emphasized that NMA is an important technique in 5G, and introduced some candidate NMA technologies including NCMA, SCMA and PDMA.

In the "5G Wireless Access: Requirement Concept and Technology" white paper published in September 2014, NTT DoCoMo introduced the technical principles and performance advantages of NOMA, which is based on the power domain, and pointed out that NOMA can be used in future 5G mobile communication networks. Furthermore, NTT DoCoMo took part in 3GPP standardization research and promotion on NMA, submitted NOMA simulation results in RAN WG1 #84bis, #85 and #86 meetings, and gave out their own suggestions on the use of NMA in mMTC and enhanced Mobile BroadBand (eMBB) scenarios.

As the leading telecom company in the USA, Qualcomm joined the research on NMA and proposed the RSMA scheme. In the "5G Waveform and Multiple Access Techniques" technical report published on November 4, 2015, Qualcomm mentioned that RSMA could satisfy asynchronous transmission and grant-free access use cases. Qualcomm is now developing 5G key techniques across three areas. In the unified radio design based on optimized OFDMA waveform, Qualcomm hopes to satisfy the requirements of specific user cases with the help of RSMA. Moreover, Qualcomm took part in the 3GPP research on NMA, and submitted proposals discussing problems such as RSMA simulation evaluation results and requirements of NMA in the 3GPP RAN WG1 #84bis, #85 and #86 meetings.

Nokia also joined the research of NMA positively, which offered some suggestions about uplink competitive multiple access with Shanghai Bell in the 3GPP RAN WG1 #84bis meeting. In the 3GPP RAN WG1 #85 meeting, Nokia proposed NOCA and IDMA schemes and gave comparisons and analyses on proposed schemes based on some aspects including receiver complexity, standardization and so on. Furthermore, Nokia took part in the NMA special topic conferences organized by IMT-2020 and submitted several proposals.

In addition to the abovementioned companies, some other companies also took part in the research of NMA. Intel proposed low code-rate spectrum spreading and frequency spectrum spreading schemes; MTK raised Repetition Division Multiple Access (RDMA) and Group Orthogonal Coded Access (GOCA); Sony Corporation gave some pieces of advice on Modulation and Coding Scheme (MCS) choice and collision in NOMA [12]; China Telecom, CMCC, FiberHome and Xinwei Telecom all participated in the research positively.

1.2.4 Current Research Status of Academia

Academia has launched research on NMA techniques including NOMA, SCMA, PDMA and MUSA. References [13, 14] introduce the advantages, chances, challenges and future research directions of NMA comprehensively and provide a good summary. Reference [13] provides a brief description to the principle of candidate NMA technologies (such as NOMA based on the power domain, SCMA, MUSA based on the code domain and so on) and some other non-orthogonal schemes (such as PDMA). This article also compares and analyzes the receiving complexity of several candidate schemes. Furthermore, to support the demands of different devices and different applications flexibly, we propose Software Defined Multiple Access (SoDeMA) scheme. To meet the different requirements of downlink and uplink communications, reference [14] presents several candidate NMA schemes separately. For downlink systems, we focus on discussing Multi-User Superposition Transmission (MUST) schemes that can be used in broadband communications and introducing the technical principle, receiver design scheme, control signaling design scheme, combination scheme with multiple antennas and performance evaluation results of three kinds of MUST schemes. For uplink systems, we introduce grant-free multiple access that could satisfy the transmission demands of massive Internet of things (IoT), such as MUSA, SCMA and so on. Since different NMA schemes apply different signal superposition methods, the research emphasis is different. Next, we will introduce the current research status of each candidate schemes separately.

NOMA allocates different transmit power to different users according to the user channel quality, then superposes user signals in the same time-frequency resource and transmits. At the receiver, SIC is used to separate signals of different users. The performance of the SIC receiver depends on the quality of receiving signals to a large extent. The more equivalent Signal-to-Noise Ratio (SNR) difference among different receiving signals, the better the performance of SIC receiver is. Aiming at downlink NOMA systems, reference [15] presents several receiver design schemes and compares their performances. Simulation results show that Codeword-level SIC (CW-SIC) receiver can approach the Block Error Rate (BLER) performance of ideal SIC receiver, while the BLER performance of Symbol-level SIC (SL-SIC) receiver is worse than CW-SIC receiver and can be affected by user transmit power easily. However, since the SL-SIC receiver does not execute encoding and decoding to the first demodulated users, its complexity is lower than CW-SIC receiver. Second, to ensure effective interference cancellation at the receiver, reasonable user pairing and power allocation are needed at the transmitter. References [16–18] propose several power allocation schemes, such as Full Search Power Allocation (FSPA) [16], statistical-based power allocation [17], and Fractional Transmit Power Allocation (FTPA) [18]. FSPA does an overall search for all user pairs and transmit power allocation. In other words, it considers all possible power allocation combinations to all candidate user pairs [16]. This algorithm can achieve better performance; however, it increases the compute complexity greatly. The statistical-based power allocation scheme proposed by reference [17] uses binary search to adjust the base station transmit power that allocated to the user with good channel quality (denoted by user 2) constantly. Based on that, it computes the restrained ergodic capacity of the user with bad channel quality (denoted by user 1). The ergodic capacity of user 1 is determined by the covariance matrix of user 1 and user 2's transmitted signals [17]. Based on these computations, the approximate expression of user 1's ergodic capacity can be achieved by using the lower limit, and the sub-optimal power allocation scheme with low complexity can be obtained. FTPA is another sub-optimal power allocation scheme based on the power control method in Long-Term Evolution (LTE) uplink. When the fairness factor equals zero, base station allocates power to each candidate user equally. With the increase of the fairness factor, power will be allocated to users with low channel gain first [18]. Reference [19] presents three corrected fractional power allocation schemes based on reference [18], which are minimum SNR-based correction, average SNR-based correction and maximum SNR-based correction. With regard to user pairing, references [20, 21] introduce two pairing schemes, one of which is the pairing between user with the best channel quality and the worst channel quality, another is the pairing between user with the best channel quality and the second best channel quality. Moreover, there are many references evaluating the performance of NOMA. References [22, 23] work on the interrupt performance and system sum data rate of the NOMA uplink and downlink, respectively. Simulation results indicate that by using reasonable power allocation method, NOMA can obtain better interrupt performance and ergodic sum data rate compared with orthogonal multiple access system. References [24–26] evaluate NOMA system performance, and discuss the NOMA throughput gain compared with orthogonal multiple access systems in different scheduled user number and different cell environment situations. The simulation results show that NOMA can improve cell throughput and cell-edge user throughput. If the cell throughput requirement can be relaxed slightly, cell-edge user throughput can be improved. With the increase of scheduled users, the throughput gain can be further improved due to the multi-user diversity gain. There is some other research on the combination of NOMA and MIMO. Reference [27] suggests that cancelling the interference among user groups by using random beamforming can increase total throughput and cell-edge user throughput; reference [28] improves this method by using random beamforming with block alignment to eliminate interference among user groups; in addition, reference [29] proposes signal alignment aided zero forcing algorithm to cancel interference among user groups and improves system throughput.

SCMA distinguishes different users by allocating a specific codebook to each user; thus, codebook design is a key factor for the SCMA system to achieve good performance and flexibility. Reference [30] gives a detailed description of the codebook design of SCMA. First, designing a multidimensional constellation with good Euclidean Distance to act as the basic constellation; then, rotating constellation to obtain an appropriate distance; finally, constructing multiple sparse codebook with different layers based on different operations (such as phase rotation). Reference [31] presents an SCMA codebook design scheme based on the Gray constellation design principle. Simulation results indicate that the BLER of SCMA is lower than Low Density Signature (LDS) and OFDMA in both Additive White Gaussian Noise (AWGN) channels and fading channels. Furthermore, SCMA can obtain a better system valid throughput than LDS and OFDMA. Reference [32] proposes a parallel SCMA system, which could reduce the decoding complexity at the receiver by transforming SCMA encoding and decoding of long codeword to SCMA encoding and decoding of short codeword. Simulation results show that the parallel SCMA system proposed by the reference can greatly reduce receiving complexity while maintaining high throughput and low BER of the system. In addition, multi-user detection algorithm of SCMA is one of the focuses for research. Reference [33] puts forward Turbo Message Passing Algorithm (MPA) receiver, which is composed of MPA algorithm and Turbo decoding. The receiver can improve the decoding performance of the system, especially in high load situations. Reference [34] presents a fractional marginalization detection algorithm, which could achieve the BER performance of MPA algorithm with a low complexity. Reference [35] proposes a weighted MPA algorithm, which determines the probability of codeword in received superposed signals by introducing weight factor, and could decrease the computing time efficiently. Reference [36] suggests simplifying SCMA detection algorithm by making a priori judgement to users with high confidence. There are many other references discussing the interrupt performance and energy efficiency of SCMA. Reference [37] evaluates the performance of uplink competitive based SCMA system. Simulation results indicate that with a given system interrupt probability, the number of active users SCMA can support 2.8 times as many as OFDMA system. Reference [38] designs the appropriate user pairing algorithm, power allocation scheme and scheduling algorithm to make SCMA effectively used in downlink transmission. Simulation results show that SCMA can improve overall system throughput compared with OFDMA. Reference [39] analyzes the energy efficiency of SCMA, and tests through system and platform. Test results indicate that SCMA can provide extra multiple access capacity and energy efficiency.

Besides, there are many other references about NMA technologies, such as PDMA, MUSA and so on. Reference [40] introduces the basic principle, transmitter and receiver design schemes and preliminary performance evaluation results of PDMA. Simulations results show that PDMA can obtain more than two times throughput gain in the uplink and more than 50% throughput gain in the downlink compared with traditional orthogonal multiple access technologies. Based on reference [40], reference [41] introduces the technical principle of PDMA, design schemes of the pattern matrix and multi-user detection technique at the receiver in a more detailed way and evaluates the performance of PDMA. System-level simulation results show that access number PDMA can support which is six times of OFDMA, and the spectrum efficiency can improve at least by 30%. Moreover, there is a lot of further research on the PDMA detection algorithms. Reference [42] proposes an iterative receiver based on minimize mean square error (MMSE) and SIC, which could control the propagation of bit error effectively and improve the system decoding performance. Reference [43] presents an advanced iterative detection receiver, which further improves the system decoding performance by implementing information iteration through decoder and detector. Reference [44] introduces a basic principle and performance evaluation results of MUSA, which employs innovative designed complex field poly-cell code and advanced SIC-based multi-user detection, which causes that, in the same time-frequency resource, the number of access users is several times that of 4G. With the help of grant-free scheme, MUSA simplifies the processes including synchronization and power control, then the realization of UE is simplified, and the UE power consumption is decreased. All the above researches indicate that compared with orthogonal access, the NMA can improve system capacity and cell-edge user capacity.

References

[1]   Goldsmith A. Wireless communications[M]. Cambridge University Press, 2005.
[2]   R1-163656. WF on multiple access for NR[R]. CMCC, Huawei, HiSilicon, Fujitsu, CATT, China Telecom, 3GPPTSG RAN WGI Meeting#84 bits, Busan, Korea, 2016.
[3]   R1-168427. WF on UL LLS for MA[R]. Huawei, HiSilicon, CATR, CATT, Spreadtrum, Fujitsu, CMCC, InterDigital, China Telecom, 3GPP TSG RAN WG1 Meeting #86, Gothenburg, Sweden, 2016.
[4]   非正交多址接入技术——5G 新助力[EB/OL]. http://www.future-forum.org/2009cn/onews.asp?id=5132.
[5]   White paper. 5G SIG Whitepaper v2.0 v1[Z]. Future Forum, 2015.
[6]   朱晨鸣, 王强, 李新, 等. 5G: 2020后的移动通信[M]. 北京: 人民邮电出版社, 2016.
[7]   Monserrat J F, Mange G, Braun V, et al. METIS research advances towards the 5G mobile and wireless system definition[J]. EURASIP Journal on Wireless Communications and Networking, 2015, (1): 53.
[8]   NET World 2020[Z]. Joint Whitepaper, final version, 2014.
[9]   IMT2020_TECH_NMA_15029_Grant-free concept and categories of MA_ZTE[Z]. 2015.
[10]   IMT2020_TECH_NMA_15030_Receiver design for grant-free MUSA_ZTE[Z]. 2015.

[11] IMT2020_TECH_NMA_15031_Simulation parameters for MA_ ZTE[Z]. 2015.

[12] R1-166651. Non-orthogonal multiple access for uplink[R]. Sony, 3GPP TSG RAN WGI Meeting#86, Gothenburg, Sweden, 2016.

[13] Dai L, Wang B, Yuan Y, et al. Non-orthogonal multiple access for 5G: solutions, challenges, opportunities, and future research trends[J]. IEEE Communications Magazine, 2015, 53(9): 74–81.

[14] Yuan Y, Yuan Z, Yu G, et al. Non-orthogonal transmission technology in LTE evolution[J]. IEEE Communications Magazine, 2016, 54(7): 68–74.

[15] Yan C, Harada A, Benjebbour A, et al. Receiver design for downlink non-orthogonal multiple access (NOMA)[C]. IEEE VTC Spring, 2015: 1–6.

[16] Benjebbour A, Saito Y, Kishiyama Y, et al. Concept and practical considerations of non-orthogonal multiple access (NOMA) for future radio access[C]. ISPACS, 2013: 770–774.

[17] Sun Q, Han S, Chin-Lin I, et al. On the ergodic capacity of MIMO NOMA systems[J]. IEEE Wireless Communications Letters, 2015, 4(4): 405–408.

[18] Saito Y, Benjebbour A, Kishiyama Y, et al. System-level performance evaluation of downlink non-orthogonal multiple access (NOMA)[C]. Proc. IEEE 24th PIMRC, 2013: 611–615.

[19] Park S, Cho D. Random linear network coding based on non-orthogonal multiple access in wireless networks[J]. IEEE Communication Letters, 2015, 19(7): 1273–1276.

[20] Ding Z, Fan P, Poor H V. Impact of user pairing on 5G non-orthogonal multiple access downlink transmissions[J]. IEEE Transactions on Vehicular Technology, 2015, (99): 1–1.

[21] Ding Z, Adachi F, Poor H V. The application of MIMO to non-orthogonal multiple access[J]. IEEE Transactions on Communication, 2016, 15(1): 537–552.

[22] Zhang N, Wang J, Kang G, et al. Uplink non-orthogonal multiple access in 5G systems[J]. IEEE Communications Letters, 2016, 20(3): 458–461.

[23] Ding Z, Yang Z, Fan P, et al. On the performance of non-orthogonal multiple access in 5G systems with randomly deployed users[J]. IEEE Signal Processing Letter, 2014, 21(12): 1501–1505.

[24] Saito Y, Benjebbour A, Kishiyama Y, et al. System-level performance evaluation of downlink non-orthogonal multiple access (NOMA)[C]. Proc. IEEE 24th PIMRC, 2013: 611–615.

[25] Saito Y, Benjebbour A, Kishiyama Y, et al. System-level performance of downlink non-orthogonal multiple access (NOMA) under various environments[C]. IEEE VTC Spring, 2015: 1–5.

[26] Tomida S, Higuchi K. Non-orthogonal access with SIC in cellular downlink for user fairness enhancement[C]. IEEE International Symposium Intelligent Signal Processing and Communications Systems (ISPACS), 2011: 1–6.

[27] Higuchi K, Kishiyama Y. Non-orthogonal access with random beamforming and intra-beam SIC for cellular MIMO downlink[C]. Vehicular Technology Conference (VTC fall), 2013.

[28] Nonaka N, Kishiyama Y, Higuchi K. Non-orthogonal multiple access using intra-beam superposition coding and SIC in base station cooperative MIMO cellular downlink[C]. Vehicular Technology Conference (VTC fall), 2014.

[29] Ding Z, Schober R, Poor H V. A general MIMO framework for NOMA downlink and uplink transmission based on signal alignment[J]. IEEE Transactions on Wireless Communications, 2016, 15: 4438–4454.

[30] Nikopour H, Baligh H. Sparse code multiple access[C]. PIMRC, 2013.

[31] Taherzadeh M, Nikopour H, Bayeseh A, et al. SCMA codebook design[C]. IEEE 80th VTC Fall, 2014.

[32] Han Y X, Zhang S H, Zhou W Y, et al. Enabling SCMA long code words with a parallel SCMA coding scheme[C]. WCSP, 2015.

[33] Lu L, Chen Y, Guo W, et al. Prototype for 5G new air interface technology SCMA and performance evaluation[C]. China Communication, 12(Supplement), 2015: 38–48.

[34] Mu H, Ma Z, Alhaji M, et al. A fixed low complexity message pass detector for up-link SCMA system[J]. IEEE Wireless Communications Letters, 2015, 4(6): 585–588.

[35] Wei D, Han Y, Zhang S, et al. Weighted message passing algorithm for SCMA[C]. WCSP, 2015.

[36] Xiao K, Xiao B, Zhang S, et al. Simplified multiuser detection for SCMA with sum-product algorithm[C]. WCSP, 2015.

[37] Au K, Zhang L, Nikopour H, et al. Uplink contention based SCMA for 5G radio access[C]. IEEE GLOBECOM, 2014: 1–5.

[38] Nikopour H, Yi E, Bayesteh A, et al. SCMA for downlink multiple access of 5G wireless networks[C]. IEEE Global Telecommunications Conference (GLOBECOM), 2014.

[39] Zhang S, Xu X, Lu L, et al. Sparse code multiple access: an energy efficient uplink approach for 5G wireless systems[C]. IEEE Global Telecommunications Conference (GLOBECOM), 2015.

[40] 康绍莉, 戴晓明, 任斌. 面向 5G 的 PDMA 图样分割多址接入 技术[J]. 电信网技术, 2015, (5).

[41] Chen S, Ren B, Gao Q, et al. Pattern division multiple access (PDMA)-a novel non-orthogonal multiple access for 5G radio networks[C]. IEEE Trans. Veh. Tech., 2016.

[42] Kong D, Zeng J, Su X, et al. Multiuser detection algorithm for PDMA uplink system based on SIC and MMSE[C]. IEEE ICCC, 2016.

[43] Ren B, Yue X, Tang W, et al. Advanced IDD receiver for PDMA uplink system[C]. IEEE ICCC, 2016.

[44] 袁志峰, 郁光辉, 李卫敏. 面向 5G 的 MUSA 多用户共享接入[J]. 电信网技术, 2015, (5).

Chapter 2
Theoretical Basis of NMA Technology

2.1 Outline of NMA Technology

Lin Liang

In the 3GPP RAN WG1 #86 meeting, there are as many as 15 NMA schemes in the proposals provided by different companies [1]. By non-orthogonal superposition of multi-user signals in physical resources through different domains (such as code domain, power domain, interleave domain and so on), these schemes break the traditional orthogonal resource allocation pattern of OMA and realize more user multiplexing in limited resources. Therefore, NMA can not only improve the system spectrum efficiency but also increase the system access capacity. Besides, by using grant free, NMA can simplify signaling procedure and decrease the radio transmission latency. Compared with OMA, NMA can approach the multi-user capacity bound, support overload transmission, realize reliable low-latency grant-free transmission, open-loop multi-user multiplexing, coordinated multi-point, and support multi-service multiplexing flexibly. Different NMA technologies can apply a unified implementation framework, while they can distinguish from each other through different resource mapping methods. In this way, different technologies can be changed smartly and related modules can be reused, which means resource utilization rate increasing and commercialization cost decreasing. Figures 2.1 and 2.2 are the unified implementation block diagrams for uplink and downlink multiple access, respectively [2].

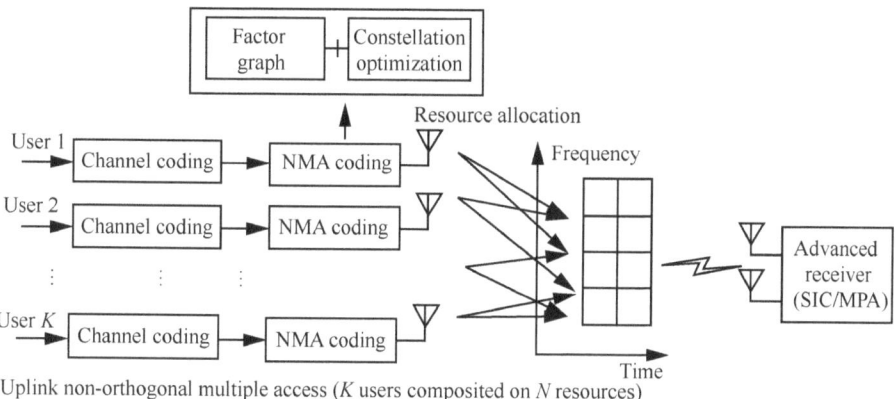

Uplink non-orthogonal multiple access (K users composited on N resources)

Figure 2.1: Unified implantation block diagram for uplink multiple access.

https://doi.org/10.1515/9783110666366-002

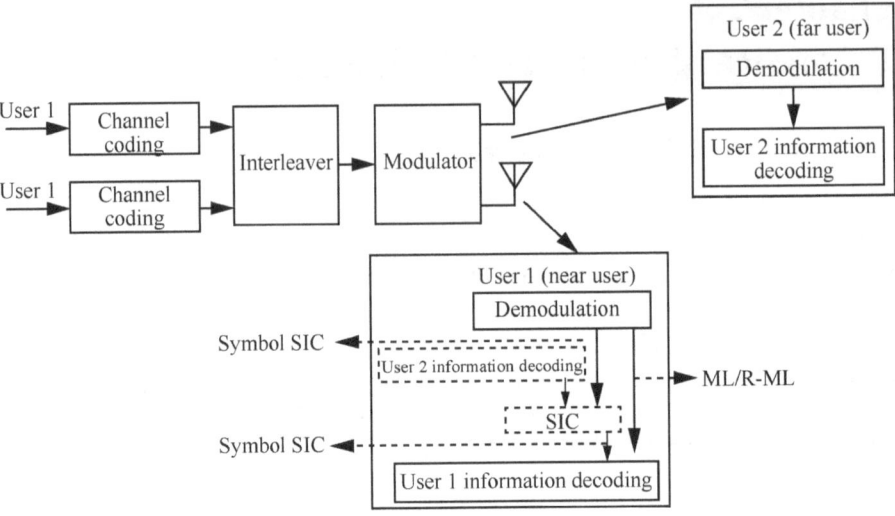

Figure 2.2: Unified implantation block diagram for downlink multiple access.

In the 3GPP RAN WG1 #85 meeting and the tenth IMT-2020 (5G) NMA special topic conference, ZTE suggested to classify current NMA technologies according to whether the user data needs spectrum spreading or not; China Telecom suggested to classify NMA technologies according to the signature sequence design of different users in the 3GPP RAN WG1 #86 meeting; In the 3GPP RAN WG1 #85 meeting, Nokia suggested to classify NMA technologies according to the receiver structure. In addition, companies such as Huawei and CATT also proposed some classification suggestions. Next, we use the classification method proposed by China Telecom in the RAN WG1 #86 meeting as an example to summarize current proposed NMA schemes. According to the signature sequence design of different users, NMA technologies can be classified into three categories: scrambling code-based NMA, interleaving-based NMA and spectrum spreading-based NMA [3]. Specific characteristics are listed in Table 2.1.

Table 2.1: Main characteristics of three kinds of NMA technologies.

	Scrambling code based	Interleaving based	Spectrum spreading based
Transmitter characteristic	Distinguish users by applying different scrambling codes	Distinguish users by applying different interleavers	Distinguish users by applying different spreading codes
Receiving algorithm	SIC	(iterative) ESE or MAP/MPA	MPA or SIC/PIC

Candidate NMA schemes based on scrambling code mainly include NOMA, RSMA and Low code rate and Signature-based Shared Access (LSSA). In NOMA [4], different user data are superposed and transmitted on the same resource based on different scrambling codes. By using SIC at the receiver, RSMA [5] is a non-orthogonal multiple access scheme combining low rate channel coding and scrambling codes (or interleavers). Users in RSMA are distinguished by scrambling codes (or interleavers) with good correlation characteristics, and advanced receiver (such as SIC receiver) is applied at the receiver to detect multi-user signals. In LSSA [6], user data are bit-level or symbol-level multiplexed based on a unique characteristic pattern for each user. The signature sequence pattern of each user is unknown to other users.

Candidate NMA schemes based on interleaving mainly include IDMA and IGMA. IDMA [7] distinguishes users by assigning different bit-level interleaver to each user. It applies iterative Elementary Signal Estimator (ESE) at the receiver to demodulate multi-user information. IGMA [8] distinguishes users by choosing different bit-level interleavers and (or) lattice mapping patterns. It applies ESE or Chip-by-Chip (CBC) Maximum A Posteriori (MAP) at the receiver to demodulate multi-user information.

Candidate NMA schemes based on spectrum spreading can be further divided into LDS code based and non-LDS code based. LDS code-based schemes include SCMA, PDMA and Low Density Spreading with Signature Vector Extension (LDS-SVE). By mapping user's coding bits directly to multidimensional codeword in complex field based on specific codebook, and then by resource mapping, SCMA [9] can realize non-orthogonal superposed transmission on same resource block in a sparse spectrum spreading way for codewords from different users. By using multi-user characteristic pattern matrix with nonequivalent diversity degree, PDMA [10] completes superposition and transmission of non-orthogonal signals from multidimension, such as time-frequency domain, power domain and space domain. At the receiver, semi-optimal multi-user detection algorithm, such as Belief Propagation (BP) and MPA, is used to detect multi-user information. LDS-SVE [11] extends signature sequences with LDS codebook. The schemes based on non-LDS code mainly include MUSA,NOCA, NCMA and Frequency Domain Spreading (FDS). MUSA [12] carries out spectrum spreading to symbols of each user with different complex spreading sequences, then superposes symbol sequences after spectrum spreading on same time-frequency resource and transmits. At the receiver, codeword-level SIC is used to separate multi-user signals. NOCA [13] uses low correlation sequences defined in LTE as spreading codeword and superposes symbol sequences after spectrum spreading on the same time-frequency resource and transmits. At the receiver, SIC is used to implement multi-user demodulation. The spreading codeword of NCMA [14] is generated from Grassman manifold. At the receiver, NCMA applies Parallel Interference Cancellation (PIC) to demodulate multi-user signals. FDS adopts spectrum spreading to modulated symbols directly. Data is transmitted on non-orthogonal time-frequency resources after spreading [15].

Moreover, many companies compare these NMA schemes. Reference [16] compares some NMA schemes from the following aspects shown in Table 2.2 which gives the comparison results of uplink NMA schemes.

Table 2.2: Comparison of uplink NMA schemes.

Scheme	How to distinguish users	Support competitive access or not	Load extend ability	Receiver complexity of base station	Standardization problem
SCMA	Codebook	Yes	Hard	High	Need to define new sparse codebook, introduce transmitter structure with joint modulation and spreading
RSMA	Sequence/ scrambling code	Yes	Easy	Medium	Need to define new waveform, new interleaver and spreading sequence
MUSA	Sequence	Yes	Easy	Medium	Need to define new short spreading sequence
NOMA	Power	No	Hard	Medium	Need to introduce new power control command
NCMA	Sequence/ scrambling code	Yes	Easy	Medium	Need to define new spreading sequence
PDMA	Pattern matrix	Yes	Hard	High	Need to define codebook matrix for spectrum spreading
IDMA	interleaver	Yes	Easy	High	Need to define new interleaver
NOCA	Sequence	Yes	Easy	Medium	Need to define low correlation sequence for spectrum spreading data symbol
IGMA	Interleaver/ lattice mapping pattern	Yes	Easy	High	Need to define new interleaver and lattice mapping pattern
FDS	Sequence	Yes	Easy	Medium	Need to define new sequence and determine proper sequence allocation Scheme
LSSA	Characteristic pattern	Yes	Easy	Medium	Need to define new characteristic pattern
LDS-SVE	RE mapping/ user eigenvector extension	Yes	Easy	High	Need to define LDS codebook, new RE mapping scheme and signature sequence vector extend means

2.2 Capacity Analysis of NMA Technology

Jie Zeng

Due to the asymmetry of communication systems, the system model of uplink and downlink are significantly different. The transmission in uplink communication systems is usually multi-point transmitting and single point receiving, with limited power of single user. But the more the transmit users are, the higher the total transmit power is. In the transmissions of uplink, joint processing is hard at the transmitter but convenient at the receiver. The corresponding uplink system communication model is called Multiple Access Channel (MAC). Downlink communication systems are usually single point transmitting and multi-point receiving with limited total transmit power. Thus, the more the simultaneous receiving users are, the less power allocated to each user is. In the transmissions of downlink, joint processing is convenient at the transmitter but hard at the receiver. The corresponding model is called Broadcast Channel (BC). Since uplink system and downlink system have different models and characteristics, their channel capacities and optimal transmission schemes are not the same. In this section, we analyze the uplink and downlink channel capacities of non-orthogonal multiple access theoretically [17].

2.2.1 Capacity Analysis of Uplink Multiple Access Channels

Suppose there are K users in the uplink multiple access channel. The maximum transmit power of user k is P_k, and the channel gain of user k is g_k. The channel bandwidth is B. Suppose the power spectral density of AWGN at the receiver is $\frac{N_0}{2}$, then the capacity bound of the K users can be expressed as:

$$\sum_{k \in S} R_k \leq B \log \left(1 + \frac{\sum_{k \in S} g_k P_k}{N_0 B} \right) \tag{2.1}$$

where $S \subset \{1, \cdots, K\}$.

Taking two users as an example, from Eq. (2.1), the rate pair (R_1, R_2) of the two users satisfies the following inequality:

$$R_1 \leq B \log \left(1 + \frac{g_1 P_1}{N_0 B} \right) \tag{2.2}$$

$$R_2 \leq B \log \left(1 + \frac{g_2 P_2}{N_0 B} \right) \tag{2.3}$$

$$R_1 + R_2 \leq B \log \left(1 + \frac{g_1 P_1 + g_2 P_2}{N_0 B} \right) \tag{2.4}$$

Equations (2.2) and (2.3) indicate that the rate of one user cannot exceed the signal-user capacity bound. Equation (2.4) indicates the sum rate of the two users that cannot exceed the signal-user capacity bound where the single-user's power equals the sum power of the two users. The polyline in Figure 2.3 gives the capacity bound for two-user AWGN channels. In Figure 2.3, the point U and the point V represent the channel capacity of user 1 and user 2 in the situation of monopolizing the resource, respectively. Meanwhile, it can be seen from Figure 2.3 that at the point A, user 1 achieves the single-user capacity bound, while the rate of user 2 is not 0 at the same time, which can be expressed as:

$$R_2^* = B\log\left(1 + \frac{g_1 P_1 + g_2 P_2}{N_0 B}\right) - B\log\left(1 + \frac{g_1 P_1}{N_0 B}\right) = B\log\left(1 + \frac{g_2 P_2}{g_1 P_1 + N_0 B}\right) \tag{2.5}$$

Then how to achieve the capacity at the point A? Two users transmit their own signals on same time-frequency resource at the transmitter, and the signals superpose in the air interface. At the receiver, SIC is used to achieve the capacity at the point A: First, decode the signal of user 2 by considering the signal of user 1 as interference; then, reconstruct the signal of user 2, delete the reconstruct signal from received signal, and decode the signal of user 1. After that, user 1 could achieve the single-user capacity bound. Similarly, the capacity at the point B could be achieved by changing the decoding order.

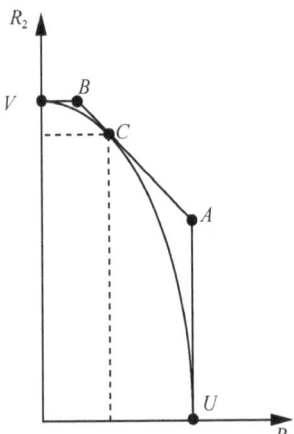

Figure 2.3: Two-user multiple access capacity bound with unequal power.

When orthogonal multiple access schemes are used, if we consider a two-user orthogonal access scheme that allocates degree of freedom with ratio α to user 1, the two-user capacity bound can be expressed as:

$$R_1 = \alpha B\log\left(1 + \frac{g_1 P_1}{\alpha N_0 B}\right) \tag{2.6}$$

$$R_2 = (1-\alpha)B\log\left(1 + \frac{g_2 P_2}{(1-\alpha)N_0 B}\right) \tag{2.7}$$

where $\alpha \in [0,1]$. The two-user capacity bound in orthogonal access schemes is illustrated by the curve in Figure 2.3. It can be seen in Figure 2.3 that non-orthogonal multiple access schemes can achieve the capacity bound, while orthogonal multiple access schemes can only reach the capacity bound at the C point, where $R_1 << R_2$ and the fairness is poor between users.

LTE employs OMA technique and needs to consider other factors such as inter-cell interference. Thus, it can only achieve the capacity bound illustrated by the curve. If NOMA is introduced in 5G systems, spectrum efficiency can be improved theoretically. On the other hand, although the best transmission scheme from the viewpoint of uplink multiple access channel is full power transmission for all users at the same time, the cellular communication system is actually a complicated network with interference, where interference cannot be fully eliminated. Massive non-orthogonal transmission will bring interference to neighbor cells. Thus, for actual performance with simultaneous transmission of multi-users, system design and engineering constraint need to be considered. Meanwhile, overall evaluation and optimization need to be carried on.

2.2.2 Capacity Analysis of Downlink Broadcast Channels

A downlink transmission model with K users can be expressed as $y_k(t) = h_k x(t) + w_k(t)$, $k = 1, 2, \cdots, K$, where $y_k(t)$ is the received signal by user k at time t; h_k is the channel coefficient of user k; $x(t)$ is the transmitted signal after joint coding by the K users at time t satisfying the power constraint $E[x^2(t)] \leq P$; $w_k(t) \sim CN(0, N_0)$ is the independently identically distributed complex Gaussian noise. Suppose $|h_1| \leq |h_2| \leq \cdots \leq |h_K|$, then the capacity region bound of downlink AWGN channel can be expressed as:

$$R_k = \log\left(1 + \frac{P_k |h_k|^2}{N_0 + \left(\sum\limits_{j=k+1}^{K} P_j\right)|h_k|^2}\right) \tag{2.8}$$

where P_k is the allocated power for user k. The capacity bound above can be achieved by superposition coding and SIC receiver.

Taking two users as an example. Suppose $|h_1| < |h_2|$, when superposition coding scheme is used, transmit signal is the superposition of the two user's signals. At the receiver, User 1 decodes its signal by considering the signal of user 2 as noise, while user 2 decodes the signal of user 1 first, then subtracts user 1's signal from the received signal and decodes. From Eq. (2.8), the achievable rate pair can be expressed as:

$$R_1 = \log\left(1 + \frac{P_1|h_1|^2}{P_2|h_1|^2 + N_0}\right) \tag{2.9}$$

$$R_2 = \log\left(1 + \frac{P_2|h_2|^2}{N_0}\right) \tag{2.10}$$

When orthogonal access scheme is used, the achievable rate pair can be expressed as:

$$R_1 = \alpha \log\left(1 + \frac{P_1|h_1|^2}{\alpha N_0}\right) \tag{2.11}$$

$$R_2 = (1-\alpha) \log\left(1 + \frac{P_2|h_2|^2}{(1-\alpha)N_0}\right) \tag{2.12}$$

where $\alpha \in [0,1]$ is the ratio of degree of freedom allocated to user 1.

Figure 2.4 gives the two-user capacity bound in downlink AWGN channels. The full line and the dotted line in the figure represent the capacity bound of the superposition coding scheme and the capacity bound of the orthogonal scheme, respectively. From Figure 2.4, it can be seen that NOMA scheme is better than orthogonal multiple access scheme except at the point A and the point B (only one user can communicate). For each achievable rate pair in orthogonal multiple access schemes, there exists a power allocation scheme, which could be used in NOMA schemes to obtain an equal or a better rate pair.

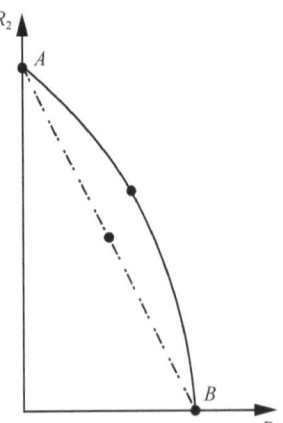

Figure 2.4: Two-user capacity bound in downlink AWGN channels.

2.3 Transmitter Design of NMA Technologies

Jie Zeng

2.3.1 Air Interface Design

1. Downlink Non-orthogonal Multiple Access

(1) User Pairing and Power Allocation

The widely used scheduling algorithms include: Max C/I algorithm, Round Robin algo-rithm and Proportional Fairness (PF) algorithm. Max C/I algorithm chooses the user with maximum C/I in the scheduling, which can guarantee maximum cell throughput without considering fairness among users. Round Robin algorithm schedules every user in turn. Therefore, it is the fairest algorithm. However, the throughput of Round Robin algorithm is low. PF algorithm uses time sliding window as a metric parameter in sched-uling and can achieve a tradeoff between system throughput and user fairness [18].

In downlink NOMA, after user pairing, the scheduler of the base station trans-mits paired users in each sub-band simultaneously. To determine the paired user set and power allocation set in each sub-band, a multi-user scheduling mechanism similar to PF scheduler is given here: in all user sets U and power sets Ps, find the user set U_{max} and power set Ps_{max} that maximize the PF schedule metric [19]:

$$Q(U,Ps) = \sum_{k \in U, Ps} \left(\frac{R_s(k,U,Ps,t)}{L(k,t)} \right)$$

$$(U_{max}, Ps_{max}) = \underset{U, Ps}{\arg \max} \, Q(U,Ps)$$

(2.13)

where $Q(U,Ps)$ denotes the PF schedule metric of user set U when the power allo-cation set is Ps, which equals the sum of PF schedule metric of all users in user set U. $R_s(k,U,Ps,t)$ is the instant throughput of user k at time t in sub-band s, and $L(k,t)$ is the average throughput of user k. The number of multiplex users in each sub-band m is determined by searching all possible user sets with different sizes.

(2) User Scheduling and MCS Choosing

In LTE and LTE-A, a single MCS is chosen in all the sub-bands allocated to one single user, thus, average Signal to Interference plus Noise Ratio (SINR) of all sub-bands can be used to choose MCS. When NOMA is used, user pairing and power allocation need to be carried out in each sub-band. Due to the mismatch between MCS choosing gran-ularity (bandwidth) and power allocation granularity (sub-band), NOMA cannot fully exhibit the gain. Besides, the finer the granularity is, the larger the signaling over-head is. Therefore, three different user scheduling/pairing and MCS choosing granu-larity combinations can be considered, as illustrated in Figure 2.5.

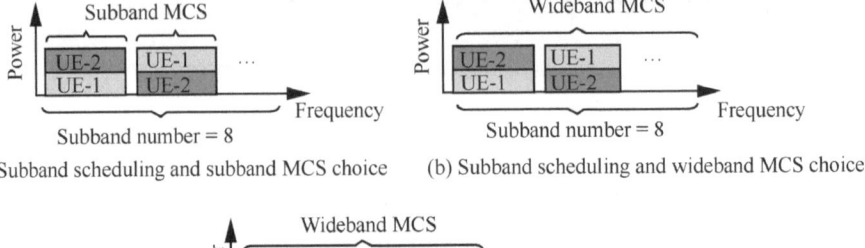

(a) Subband scheduling and subband MCS choice (b) Subband scheduling and wideband MCS choice

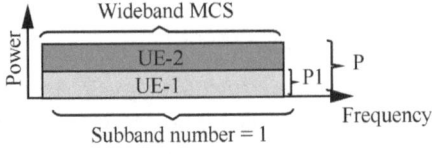

(c) Wideband scheduling and wideband MCS choice

Figure 2.5: User scheduling and MCS choosing granularity in NOMA.

2. Uplink Non-orthogonal Multiple Access

There are some commonalities between the uplink NOMA air interface design and downlink NOMA air interface design. There are also two main differences: one is the schedule design, and the other is the power control. A detailed introduction on the schedule and power control differences between uplink and downlink NOMA will be given in the following [20].

(1) Schedule Design

The grant-based and grant-free uplink NOMA schemes were widely discussed in the 3GPP RAN WG1 #84bis meeting. In the end, it is agreed that at least in uplink mMTC scenario, grant-free mechanism should be studied.

In NOMA systems, resources include multiple access physical resource and multiple access signature sequence. Multiple access physical resource includes time-frequency resource block, while multiple access signature sequence, which could be codeword, spreading sequence, interleaver and/or mapping pattern, is bound up with specific multiple access schemes. At the transmitter, different users employ different multiple access signature sequences and transmit on the same physical resource. In current LTE framework, uplink transmission is scheduling based on serving eNodeB. The orthogonal resource allocation of eNodeB can realize orthogonal multiple access for grant-based uplink transmission in LTE, therefore intra cell interference caused by resource collisions is avoided. However, the orthogonal design method has great limitations for NR in supporting the requirements of various services. On the one hand, uplink grant free NOMA can improve spectrum efficiency and system capacity. On the other hand, the latency can be decreased, which is more suitable for mMTC and Ultra-Reliable and Low Latency Communications (URLLC) scenarios [21].

To support uplink grant-free transmission, the network needs to pre-configure multiple access physical resource. Then, users can determine corresponding available multiple access physical resource. Generally, there are two multiple access signature assign methods: the users choose a multiple access signature sequence randomly; pre-define the user's multiple access signature sequence. Actually, these two methods are not mutually exclusive from the viewpoint of standardization [22].

In uplink grant-free transmissions, users can choose multiple access resource and transmit slot spontaneously. At the receiver, the base station does not know the user number superposed in one specific time-frequency resource block. Thus, preambles are needed in uplink grant-free transmissions to avoid MCS blind detection, time offset and frequency offset and decrease the decoding complexity at the receiver [23].

One function of the preamble is to identify active users. Each user chooses a preamble from a pre-defined sequence pool which has good cross-correlation features. The receiver recognizes preambles in a specific time-frequency resource by the blind detection. The collision probability (multiple users choose the same preamble) determines the supported user number. The size of the sequence pool needs to support the target connection density and packet arrival rate in mMTC scenario. However, the size of the sequence pool is related to the length of the preamble sequence. To guarantee acceptable overhead, the sequence length should not be too long. In addition, the multi-user blind detection complexity increases when the size of the sequence pool becomes larger. To summarize, the supported user number, overhead and complexity need to be considered together when defining the size of the preamble sequence pool.

In the mMTC scenario, although the receiver can obtain the preambles on specific time-frequency resource by blind detection, there may be massive access users in the same cell. It is then unrealistic to take user ID along during the transmission of preamble sequence. Unless the signal can be decoded correctly, the receiver cannot know the user ID. In grant-free transmissions, the user generally has several multiple access signature sequences, such as spreading sequence, interleaving pattern, preamble, Demodulation Reference Signal (DM-RS) and so on. To fully utilize the preamble detection and release the burden of the blind detection, the chosen preamble sequence can be seen as a temp user ID. Therefore, the mapping between preambles and multiple access signature sequences should be designed. When the size of the preamble sequence pool equals the size of the multiple access signature sequence pool, one-to-one mapping can be applied simply. When the size of the multiple access signature sequence pool is limited, multiple preamble sequences can be mapped to the same spreading sequence/code/interleaving pattern, and the collision of spreading sequences/codes/interleaving patterns can be solved by distinctions of the users in other domains (power domain/space domain).

However, the mapping between preamble and DM-RS pattern is not intuitive. In NMA systems, although multi-users can transmit data on shared resource blocks, the user number can be supported actually is often limited by the DM-RS resource. To decrease the DM-RS collision probability, increasing some overhead to allocate more resources for DM-RS can be considered.

(2) Power Control

The difference between the power control in uplink NOMA and power control in downlink NOMA is mainly reflected in two aspects. First, the limitations on optimized transmit power is different. There is only one limitation on downlink transmit power, which is the maximum transmit power of BS; while in the uplink, the limitation on transmit power is the maximum transmit power of a single user. Second, the transmit power control design method is different: in the downlink, the superposed signal user received experiences the same channel, which means the received signal for different users at the receiver has the same channel gain. Thus, the design purpose of downlink power control is to create user-level difference in power domain artificially, so as to separate user signals like SIC; while in the uplink, since signals from different users experience different channels, the received power is different. Besides, when NOMA is used in the uplink, simultaneous transmission of multiple UEs may cause drastically Inter-Cell Interference (ICI) increase. But in the downlink, no matter how many multiplexed users are, the maximum transmission power of BS is a fixed value, NOMA does not cause drastically ICI increase. Therefore, uplink power control includes the following two aspects: when the difference of the channel gains is not big, a difference can be created in received signals artificially; user transmit power is controlled to avoid too much ICI to neighbor cells.

Next, we take the uplink NOMA power control algorithm based on the sum of PF schedule metric maximization as an example, to explain the basic procedures of power control. First, adopt the Fractional Transmit Power Control (FTPC) method to obtain the basic transmit power of user k:

$$P_k = \min\{P_{\max}, 10\lg(M_k + P_0 + \alpha \cdot PL_k)\} \tag{2.14}$$

where $P_{\max}$ denotes the maximum transmit power for a single user; M_k is the allocated frequency resource block number; PL_k is the path loss of user k (including the path loss related with distance and normal shadow fading); P_0 is the target received power when the path loss is zero; α denotes the FTPC path loss factor. Then, we determine the total transmit power of each NOMA user set. The total transmit power for each user set U can be expressed as:

$$P_{total,U} = \beta \cdot N \cdot P_{avg} \tag{2.15}$$

$$P_{avg} = \frac{1}{K} \sum_{k=1}^{K} P_k \tag{2.16}$$

where $\beta \in [0,1]$ is the optimization parameter considering the ICI to other cells; N is the user number in the NOMA user set; P_{avg} denotes the average transmit power of users in the cell. Finally, after determining the total transmit power of each user set, exhaustive searching method same as downlink is used to choose the optimal power set $Ps_{\max}$ and the optimal user set $U_{\max}$. Noticing that the best value of β is a variant for different cells, it needs to be obtained by joint optimization of multiple cells.

The tolerance of error propagation is different between downlink and uplink NOMA. When the multiplexed user number is large, uplink NOMA will cause more serious error propagation than downlink, which should be considered when design the scheduler in actual systems.

2.3.2 Design with MIMO

MIMO is a key technique to improve the spectrum efficiency in LTE/LTE-A. By applying MIMO to NMA, the spectrum efficiency can be further improved [19]. This section takes the combination of NOMA and MIMO as an example to explain the specific scheme of combining NMA and MIMO.

1. Combination of NOMA and MIMO in Downlink

One method of the combining NOMA and MIMO in downlink is to generate multiple power levels with NOMA and transmit by employing single-user MIMO (SU-MIMO) and/or multi-user MIMO (MU-MIMO) in each power level.

Another method is to decompose MIMO channel into several SIMO channels and adopt NOMA in each SIMO channel. Figure 2.6 is a diagram of combining NOMA and 2×2 MIMO in downlink with this method (random beamforming and Interference Rejection Combining SIC (IRC-SIC) receiver are used) [19]. In this method, multiple transmit beams can be generated by opportunistic beamforming, and superposition coding is used in each transmission beam. Space domain filtering is used with multiple receiving antennas to eliminate interference inside the beam at the user side (assuming UE has multiple antennas), then SIC is employed to separate multi-user signals.

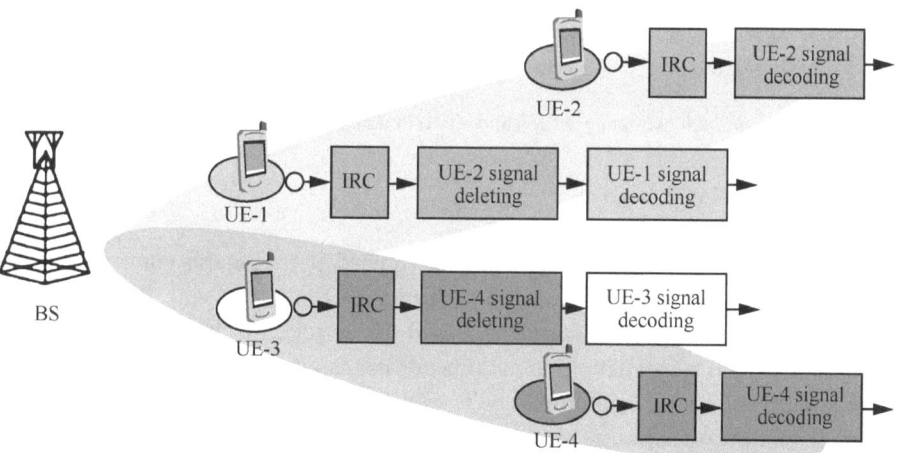

Figure 2.6: Combination diagram of NOMA and 2×2 MIMO in downlink.

2. Combination of NOMA and MIMO in Uplink

In uplink systems with combination of NOMA and MIMO, when the UE has multiple antennas, the degree of freedom in each point-to-point link is greater than 1. Therefore, each UE can transmit multi-stream data. Figure 2.7 gives the combination diagram of NOMA and 2×2 MIMO in uplink. In Figure 2.7(a), multi-user signals are separated in power domain, and the space degree of freedom is used to multiplex multiple data streams from a single user. While in Figure 2.7(b), users are distinguished from each other in power domain and space domain [19]: {UE-1} and {UE-2, UE-3} are separated in power domain, meanwhile UE-2 and UE-3 are further separated in space domain. UE-1 transmits with SU-MIMO mode, while UE-2 and UE-3 transmit with MU-MIMO mode.

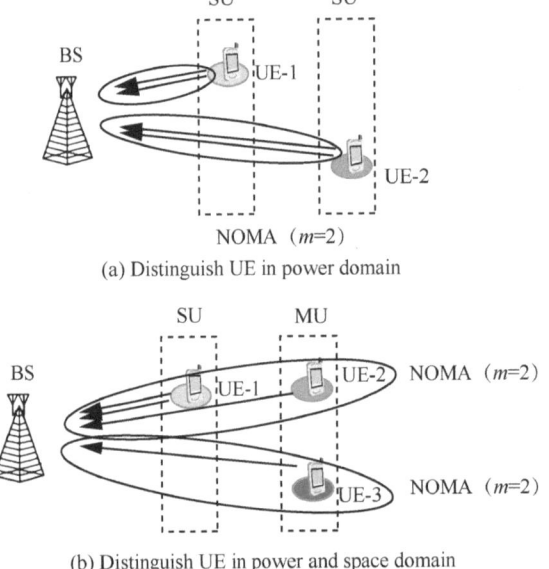

(a) Distinguish UE in power domain

(b) Distinguish UE in power and space domain

Figure 2.7: Combination diagram of NOMA and MIMO in uplink.

2.3.3 Constellation Design

In NMA, system BER performances can be optimized by reasonable constellation design, while receiver complexity can be decreased by mapping with reduced order. The purpose of the constellation design is to find a good distance criterion among multidimensional constellation points (Euclidean Distance and/or Product Distance) to maximize the code/beamforming gain. Generally speaking, the multidimensional constellation maximizing the minimum Euclidean Distance can be taken as a basic constellation. In low-rate situations, heuristic optimization algorithms can be used to obtain the basic constellation. Then different constellations are constructed by rotation (such as phase rotation). If the real part and the imaginary part are mutu-ally independent in the constructed complex constellation, the decoding com-plexity can be decreased by multidimensional constellation expansion. Figure 2.8 shows a multidimensional constellation expansion method by separating the real part and the imaginary part [24].

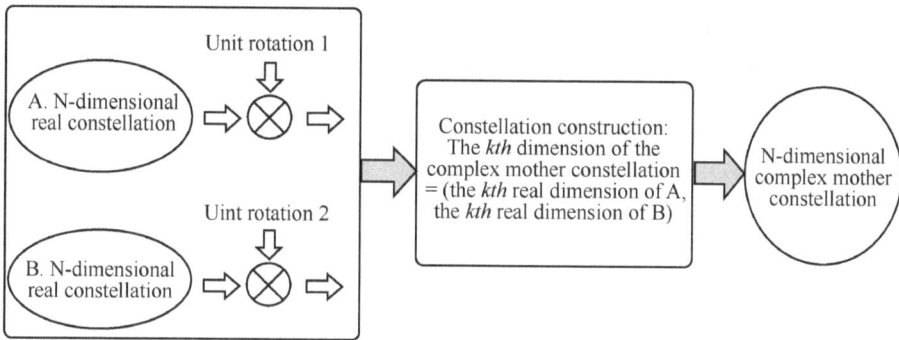

Figure 2.8: Multidimensional constellation expansion.

Figure 2.9 presents an example of obtaining a 16-point constellation by multidimen-sional constellation expansion, in which the rotation angle of maximizing the mini-mum Euclidean Distance is $\arctan\left(\frac{1+\sqrt{5}}{2}\right)$.

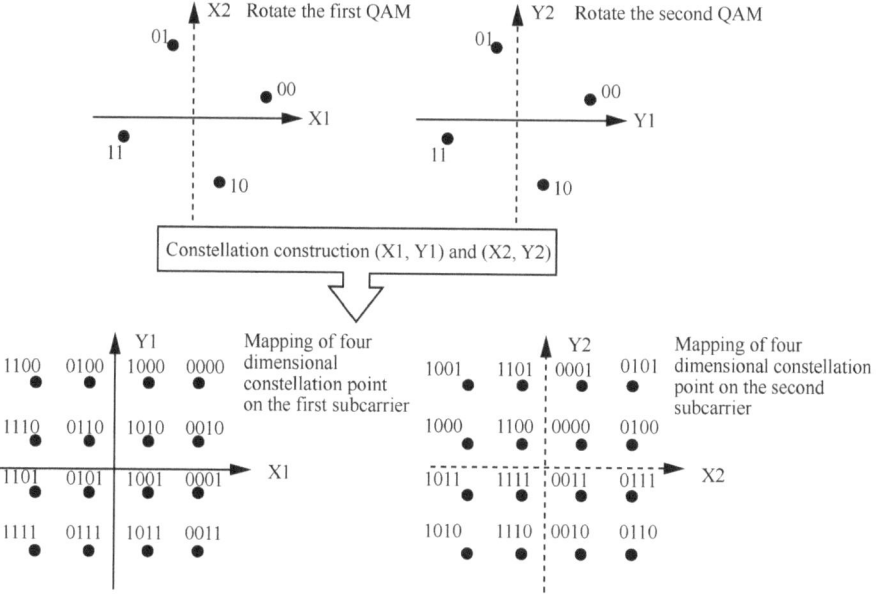

Figure 2.9: Example of constellation expansion.

2.4 Advanced Receiver Design

Jie Zeng

The superposition transmission of users in NMA can cause multi-user interference, and the detection algorithm of the receiver affects the NMA performance directly. To alleviate the multi-user interference in NMA systems, advanced receiver design should be considered. Currently, multiple receivers aiming at different multiple access technical characteristics have been proposed by the industry.

2.4.1 Optimal Receiver

The MAP receiver can achieve the best symbol-level detection performance. When a priori information of each symbol is equal, Maximum Likelihood (ML) and MAP have the same performance. Therefore, in general communication systems, ML can achieve the best symbol-level detection [25]. To find the symbol vector with the minimum Euclidean Distance, ML and MAP detection need to calculate Euclidean Distances from all possible symbol vector combinations to the received signal. The computing complexity of ML and MAP grows exponentially with the increase of the user number. Therefore, the complexity of ML and MAP is high, and hardware implementation is difficult. Besides, the codeword-level ML detection in channel coding has a better performance than symbol-level ML detection. However, its complexity is extremely high at the receiver.

2.4.2 IC Receiver

When receiving multiple signals simultaneously, Interference Cancellation (IC) receiver detects the signal from interference user first and reconstructs the interference signal and subtracts the reconstructed signal from the received signals. IC receiver can be divided into SIC receiver and PIC receiver.

1. SIC Receiver

The principle of SIC receiver is to detect and reconstruct user signal successively in a descending order of the signal power, then subtract reconstructed signal from received signals. The reconstruction can be symbol-level or codeword-level. Hence, SIC can be divided into SL-SIC and CW-SIC. In SL-SIC, the first detected user executes demodulation, hard decision and modulates again. Then, the modulated signal is subtracted from the received signal. In CW-SIC, the first detected user signal is demodulated and decoded, and the information bits derived are coded and modulated and then deleted from the received signals. Channel coding is employed in the signal-detecting procedure in CW-SIC. Therefore, compared with SL-SIC, CW-SIC

can improve the probability of correct signal recovery and decrease the error propagation. However, the complexity of CW-SIC is greatly increased due to the channel encoding/decoding [26]. Figure 2.10 provides the principle block diagram of
SIC receiver [26]. The SIC receiver applies a group of linear receivers, each of
which detects the data of the user with the strongest signal, then detected signal
is deleted from the received signal. The steps repeat until the signal detection of
all users is completed.

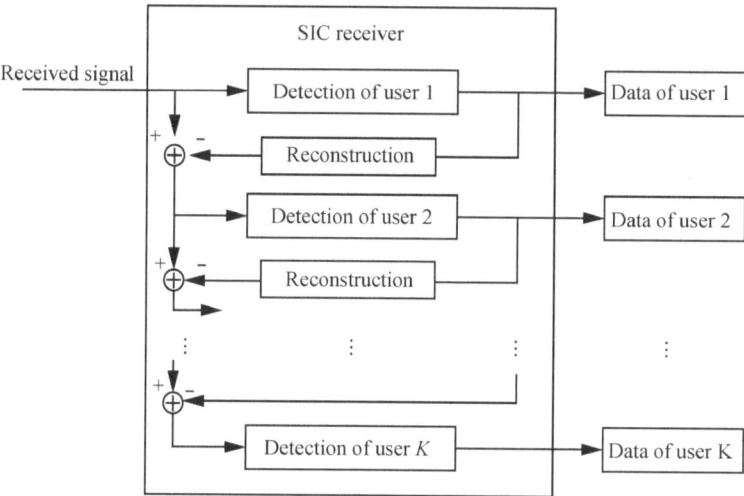

Figure 2.10: The principle block diagram of SIC receiver.

Suppose the received multi-user signal is **y**, $x_{(i)}$ denotes the ith signal that needs to
be detected, $\hat{x}_{(i)}$ denotes the estimate of $x_{(i)}$. Subtracting $\hat{x}_{(1)}$ from the received signal
to form the remainder signal in the first stage.

$$\tilde{\mathbf{y}}_{(1)} = \mathbf{y} - \mathbf{h}_{(1)}\hat{x}_{(1)} = \mathbf{h}_{(1)}\left(x_{(1)} - \hat{x}_{(1)}\right) + \mathbf{h}_{(2)}x_{(2)} + \cdots + \mathbf{h}_{(N_{tx})}x_{(N_{tx})} + z \qquad (2.17)$$

where $\mathbf{h}_{(i)}$ is the ith column vector of the channel matrix. If $x_{(1)} = \hat{x}_{(1)}$, then the interference can be deleted successfully when estimating $x_{(2)}$, otherwise, error propaga-
tion will emerge.

2. PIC Receiver

Different from SIC receiver that deletes interference in sequence, the detection, reconstruction and deletion of interference signals of all users can be carried on simultaneously in PIC, as illustrated in Figure 2.11. Thus, the processing delay of PIC
detection is lower than SIC detection. However, PIC has large amount of calculation,
and the system output may not be convergent when the user power and channel
state information are unknown.

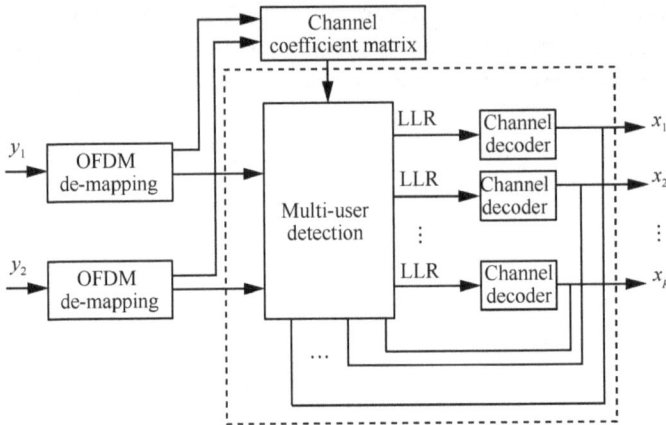

Figure 2.11: The principle block diagram of PIC detection.

2.4.3 BP Receiver

BP, namely MPA, are iterative decoding algorithms based on the factor graph. BP delivers "message" back and forth between the Variable Node (VN) and the Function Node (FN) by the side of factor graph. A stable value can be obtained after many iterations. Then, the optimal decision is made according to the iteration results. Non-orthogonal multiple access schemes based on spreading sequence can decrease the complexity of BP by utilizing the sparsity of spreading sequence, which makes BP more suitable for NOMA systems, such as PDMA, SCMA and so on.

Figure 2.12 is a factor graph containing J VNs and K FNs. ∂j denotes the FN set adjacent to VN_j, while ∂k denotes the VN set adjacent to FN_k.

$$y_k = \sum_{j\in\partial k} h_{kj}x_{kj} + n_k, \quad k = 1, \cdots, K \tag{2.18}$$

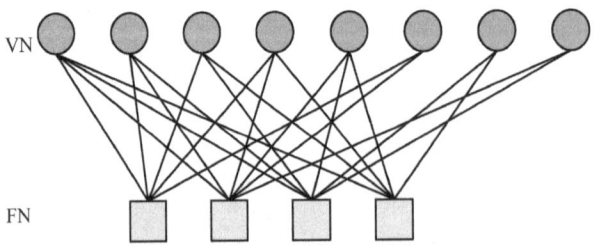

Figure 2.12: Factor graph.

In each BP iteration, "message" is first delivered from VN to FN, then FN calculates the outer information and delivers the outer information to VN. Suppose $|X^j| = M$, $\mathbf{x}_j \in X^j$. $\{V_{j \to k}^{(t)}(\mathbf{x}_j)\}$ denotes the "message" delivered from VN_j to FN_k in the t-th iteration, $\{U_{k \to j}^{(t)}(\mathbf{x}_j)\}$ denotes the "message" delivered from FN_k to VN_j in the t-th iteration. They can be calculated according to the following formulas [27]:

$$V_{j \to k}^{(t)}(\mathbf{x}_j) = P(\mathbf{x}_j) \prod_{l \in \partial j \backslash k} U_{l \to j}^{(t-1)}(\mathbf{x}_j) \tag{2.19}$$

$$U_{k \to j}^{(t)}(\mathbf{x}_j) \leftarrow \sum_{(x_p) \partial k \backslash j} \frac{1}{\pi N_0} \exp\left[\frac{1}{N_0}\left\|y_k - h_{kj}x_{kj} - \sum_{p \in \partial k \backslash j} h_{kp}x_{kp}\right\|^2\right] \prod_{p \in \partial k \backslash j} V_{p \to k}^{(t)}(\mathbf{x}_p) \tag{2.20}$$

where $\partial j \backslash k$ denotes the FN set adjacent to VN_j except FN_k; $\sum_{(x_p) \partial k \backslash j}$ denotes the sum of all $\mathbf{x}_p \in X^p$ from all $p \in \partial k \backslash j$.

When the iteration is over, for all $j = 1, \cdots, J$,

$$V_j(\mathbf{x}_j) = P(\mathbf{x}_j) \prod_{k \in \partial j} U_{k \to j}^{(t-1)}(\mathbf{x}_j) \tag{2.21}$$

2.4.4 BP-IDD Receiver

Belief Propagation based Iterative Detection and Decoding (BP-IDD) algorithm makes joint iteration of BP detection and Turbo decoding to further improve the performance of BP detection. Figure 2.13 is the principle block diagram of BP-IDD. The basic principle is obtaining the soft information that output from the decoder, then mapping the

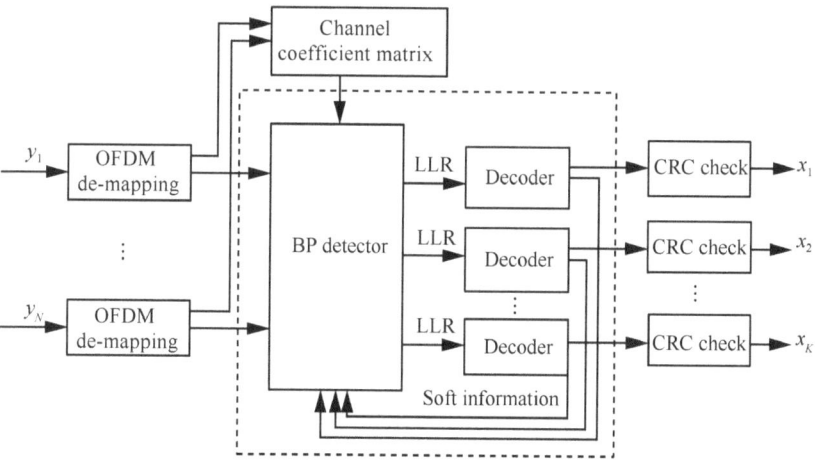

Figure 2.13: Principle block diagram of BP-IDD.

soft information from bit to symbol and utilizing the mapped symbol as the input priori information of the BP detector. The information after decoding can be fully utilized to further improve the performance of BP detecting algorithm. The BP-IDD receiver is usually used in PDMA and SCMA systems.

Figure 2.14 presents the factor graph of the BP-IDD algorithm [28], where the iteration detection between the User Node (UND, used as VN) and the Channel Node (CND, used as FN) is the iteration in BP detection. The iteration between UND and FN is the decoding procedure of Turbo. The system BLER performance is efficiently improved by the iterative decoding and detecting. Compared with BP detector, BP-IDD can support user signal separation in high load situations.

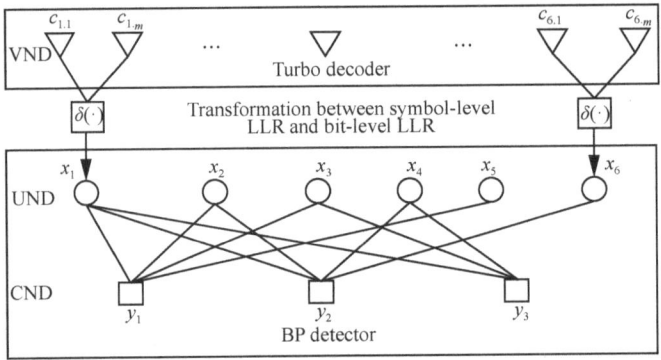

Figure 2.14: Factor graph of the BP-IDD algorithm.

SCMA applies the Turbo MPA receiver which is an outer loop receiver composed of a MAP detector and a Turbo decoder and has the same principle with BP-IDD receiver. Besides iterative detecting and decoding with a Turbo decoder, the MPA detector can also carry out iterative detecting and decoding with a Low Density Parity Check (LDPC) decoder.

2.4.5 BP-IDD-IC Receiver

Based on the characteristics of NOMA, IC technique can be introduced into BP-IDD receiver, which called the BP-IDD-IC receiver as illustrated in Figure 2.15. Compared with BP-IDD, BP-IDD-IC can achieve a better detection performance. The soft information output from the decoder is fed back to the BP detector, and the hard information output carries out Cyclic Redundancy Check (CRC). If CRC is passed, then the soft information of the user is reconstructed and subtracted from the information fed back to the BP detector. BP-IDD-IC detector can approach the detection performance of ML with an acceptable complexity.

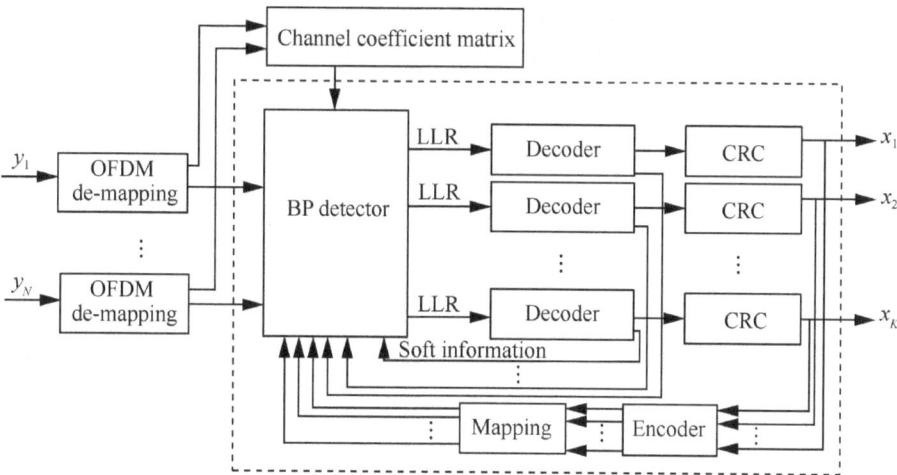

Figure 2.15: Block diagram of BP-IDD-IC receiver.

2.4.6 IS Receiver

Interference Suppression (IS) receiver uses linear filter to suppresses interference instead of eliminating it. There are three kinds of IS receivers as follows [29]:

(1) Linear Minimum Mean Square Error IRC (LMMSE-IRC)
LMMSE-IRC receiver is the baseline of the 3GPP Rel-11 MMSE-IRC research, whose advantage is not needed to know any parameters about interference information.

(2) Enhanced LMMSE-IRC (E-LMMSE-IRC)
E-LMMSE-IRC needs to consider the interference channel estimation and other parameters. Consider a main interference, then the signal model can be expressed as:

$$
\begin{aligned}
y &= H_0 x_0 + H_1 x_1 + \left(\sum_{i=2}^{K} H_i x_i \right) + n \\
&= H_0 x_0 + H_1 x_1 + z \\
&= H_0 x_0 + v
\end{aligned}
\tag{2.22}
$$

where z denotes non-principal interference and noise; v denotes overall interference and noise. E-LMMSE-IRC receiver can be expressed as:

$$
W = \hat{H}_0^H \left(\hat{H}_0 \hat{H}_0^H + \hat{H}_1 \hat{H}_1^H + \hat{R}_z \right)^{-1}
\tag{2.23}
$$

where $\hat{R}_z$ is the estimate of R_z, $R_z = E\left[zz^H \right]$.

(3) Widely LMMSE-IRC (WLMMSE-IRC)

WLMMSE-IRC enhances interference suppression by utilizing the degree of freedom of the real part and the imaginary part of the received signal. The received signal can be expressed as:

$$\mathbf{r} = \mathbf{H}_0 \mathbf{x}_0 + \mathbf{n} \tag{2.24}$$

The representation for augmented matrix is:

$$\mathbf{r}^A = \mathbf{H}_0^A \mathbf{x}_0^A + \mathbf{n}^A \tag{2.25}$$

where $\mathbf{r}^A = \begin{bmatrix} I(r) \\ Q(r) \end{bmatrix}$; $\mathbf{H}_0^A = \begin{bmatrix} I(\mathbf{H}_0) & -Q(\mathbf{H}_0) \\ Q(\mathbf{H}_0) & I(\mathbf{H}_0) \end{bmatrix}$; $\mathbf{x}_0^A = \begin{bmatrix} I(\mathbf{x}_0) \\ Q(\mathbf{x}_0) \end{bmatrix}$; $\mathbf{n}^A = \begin{bmatrix} I(\mathbf{n}) \\ Q(\mathbf{n}) \end{bmatrix}$. $I(\cdot)$ and $Q(\cdot)$ denotes taking the real part and the imaginary part of the parameter, respectively. Therefore, all the variables are real numbers. The symbol estimate of WLMMSE-IRC is:

$$\hat{\mathbf{x}}_0^A = \left(\hat{\mathbf{C}}_{rr}^A \right)^{-1} \hat{\mathbf{H}}_0^A \hat{\mathbf{C}}_{xx}^A \mathbf{r}^A = \mathbf{g} \mathbf{r}^A \tag{2.26}$$

$$\mathbf{g} = \left(\hat{\mathbf{C}}_{rr}^A \right)^{-1} \hat{\mathbf{H}}_0^A \hat{\mathbf{C}}_{xx}^A \tag{2.27}$$

where $\hat{\mathbf{C}}_{rr}^A$, $\hat{\mathbf{H}}_0^A$ and $\hat{\mathbf{C}}_{xx}^A$ are the estimate of the covariance of the received signal, the channel and the covariance of transmitted signals, respectively.

References

[1] R1-168427. WF on UL LLS for MA, Huawei, HiSilicon, CATR, CATT, Spreadtrum, Fujitsu, CMCC, Inter Digital[R]. China Telecom, 3GPP TSG RAN WG1 Meeting #86, Gothenburg, Sweden, 2016.

[2] Future Forum. 5G Whitepaper[Z]. 2015.

[3] R1-167445. Classification of candidate UL non-orthogonal MA schemes[R]. China Telecom, 3GPP TSG RAN WG1 Meeting #86, Gothenburg, Sweden, 2016.

[4] R1-165175. Initial views and evaluation results on non-orthogonal multiple access for NR[R]. NTT DOCOMO, 3GPP TSG RAN WGI Meeting #85, Nanjing, China, 2016.

[5] R1-164688. RSMA[R]. Qualcomm Incorporated, 3GPP TSG RAN WGI Meeting #85, Nanjing, China, 2016.

[6] R1-164869. Low code rate and signature based multiple access scheme for New Radio[R]. ETRI, 3GPP TSG RAN WGI Meeting #85, Nanjing, China, 2016.

[7] R1-165021. Performance of interleave division multiple access (IDMA) in combination with OFDM family waveforms[R]. Nokia, Alcatel-Lucent Shanghai Bell, 3GPP TSG RAN WGI Meeting #85, Nanjing, China, 2016.

[8] R1-163992. Non-orthogonal multiple access candidate for NR[R]. Samsung, 3GPP TSG RAN WGI Meeting #85, Nanjing, China, 2016.

[9] R1-164037. LLS results for uplink multiple access[R]. Huawei, HiSilicon, 3GPP TSG RAN WGI Meeting #85, Nanjing, China, 2016.

[10] R1-163383. Candidate solution for NMA[R]. CATT, 3GPP TSG RAN WGI Meeting #84bis, Busan, Korea, 2016.

[11]   R1-164329. Initial LLS results for UL non-orthogonal multiple access[R]. Fujitsu, 3GPP TSG RAN WGI Meeting #85, Nanjing, China, 2016.

[12]   R1-164269. Contention-based non-orthogonal multiple access for UL mMTC[R]. ZTE, 3GPP TSG RAN WGI Meeting #85, Nanjing, China, 2016.

[13]   R1-165019. Non-orthogonal multiple access for new radio[R]. Nokia, Alcatel-Lucent Shanghai Bell, 3GPP TSG RAN WGI Meeting #85, Nanjing, China, 2016.

[14]   R1-162517. Considerations on DL/UL multiple access for NR[R]. LG Electronics, 3GPP TSG RAN WGI Meeting #84bis, Busan, Korea, 2016.

[15]   R1-162385. Multiple access schemes for new radio interface[R]. Intel Corporation, 3GPP TSG RAN WGI Meeting #84bis, Busan, Korea, 2016.

[16]   R1-167247. Overview of the proposed non-orthogonal MA schemes[R]. Nokia, Alcatel-Lucent Shanghai Bell, 3GPP TSG-RAN WG1 Meeting #86, Gothenburg, Sweden, 2016.

[17]   Goldsmith A. Wireless communications[M]. Cambridge University Press, 2005.

[18]   王映民, 孙韶辉. TD-LTE 技术原理与系统设计[M]. 北京: 人民邮电出版社, 2010.

[19]   Luo F L, Zhang C. Non-orthogonal multiple access (NOMA): concept and design[C]. Signal Processing for 5G: Algorithms and Implementations, Wiley-IEEE Press eBook Chapters, 2016.

[20]   Thomas M C, Joy A T. Elements of information theory (Second Edition)[M]. Hoboken NewJersey: John Wiley & Sons, Inc., 2006.

[21]   R1-1609399. Discussion on grant-free based UL transmission[R]. Lenovo, 3GPP TSG-RAN WG1 Meeting #86bis, Lisbon, Portugal, 2016.

[22]   R1-1608757. Consideration on grant-free transmission[R]. CATT, 3GPP TSG-RAN WG1 Meeting #86bis, Lisbon, Portugal, 2016.

[23]   R1-1608955. Considerations on the preamble design for grant-free non-orthogonal MA[R]. ZTE, 3GPP TSG-RAN WG1 Meeting #86bis, Lisbon, Portugal, 2016.

[24]   Taherzadeh M, Nikopour H, Bayesteh A, et al. SCMA codebook design[C]. 2014 IEEE 80th Vehicular Technology Conference (VTC2014-Fall), 2014: 1–5.

[25]   R1-166875. Considerations on the receiver types and NOMA schemes[R]. LG Electronics, 3GPP TSG-RAN WG1 Meeting #86, Gothenburg, Sweden, 2016.

[26]   Yan C, Harada A, Benjebbour A, et al. Receiver design for downlink non-orthogonal multiple access (NOMA)[C]. VTC Spring, 2015: 1–6.

[27]   Xiao K, Xiao B, Zhang S, et al. Iterative detection and decoding for SCMA systems with LDPC codes[C]. Wireless Communications & Signal Processing (WCSP), 2015: 1–5.

[28]   Chen S, Chen B, Gao Q, et al. Pattern division multiple access (PDMA)-a novel non-orthogonal multiple access for 5G radio networks[C]. IEEE Transactions on Vehicular Technology, IEEE Early Access Articles, 2016: 3185–3196.

[29]   3GPP TR 36.866. Study on network-assisted interference cancellation and suppression (NAIC) for LTE[S]. 2014.

Chapter 3
Candidate NMA Technologies in 5G

3.1 SCMA

Jie Zeng

3.1.1 Technical Principle

SCMA is a code division non-orthogonal multiple access technology based on multidimensional modulation and sparse code spreading. Figure 3.1 is the implementation framework of uplink SCMA system. At the transmitter, LDS and multidimensional modulation are combined. The optimal codebook set is chosen by conjugation, transposition, phase rotation and so on. Users then transmit information based on allocated codebooks. The coded bits after Turbo encoder are mapped directly to multidimensional codeword in multiplex field by SCMA coding, then Physical Resource Element (PRE) mapping is used. The codewords of different users composited non-orthogonally on same resource blocks in the way of sparse spreading. At the receiver, information bits of multiple users are recovered by PRE demapping, SCMA decoding and Turbo decoding.

SCMA can be used in both uplink and downlink systems [1, 2]. In uplink systems, SCMA superposes and transmits signals from different users non-orthogonally, thereby more users can be served than LTE with the same resource number, which makes massive connections within the reach. The test results of the prototype verification system developed by Huawei shows that SCMA can increase the connected user number by 150%–300% compared with LTE, which is a 150%–300% promotion of the network overall throughput. Besides, with the help of the blind detection and the insensitive characteristic of SCMA in codeword collision, grant-free random competitive access can be implemented. Grant-free access can greatly decrease signaling overhead of the dynamic request and authorization during access. Therefore, grant-free access reduces the system implementation complexity, access latency and user power consumption. In downlink systems, SCMA can enhance link reliability and robustness by utilizing the multidimensional modulation and frequency diversities. In addition, SCMA can reduce the number of mapping constellation points by multidimensional modulation, which decreases the receiver complexity efficiently. Moreover, sparse transmission could suppress inter-layer interference, which implies support for more symbol collisions with a low receiver complexity [3].

Although SCMA has a lot of advantages, there are still many challenges. First, the implementation complexity of the current SCMA receiver is higher than 4G commercial systems. On the one hand, receiver algorithms with low complexity need to be studied; on the other hand, a reasonable tradeoff between complexity and performance needs

https://doi.org/10.1515/9783110666366-003

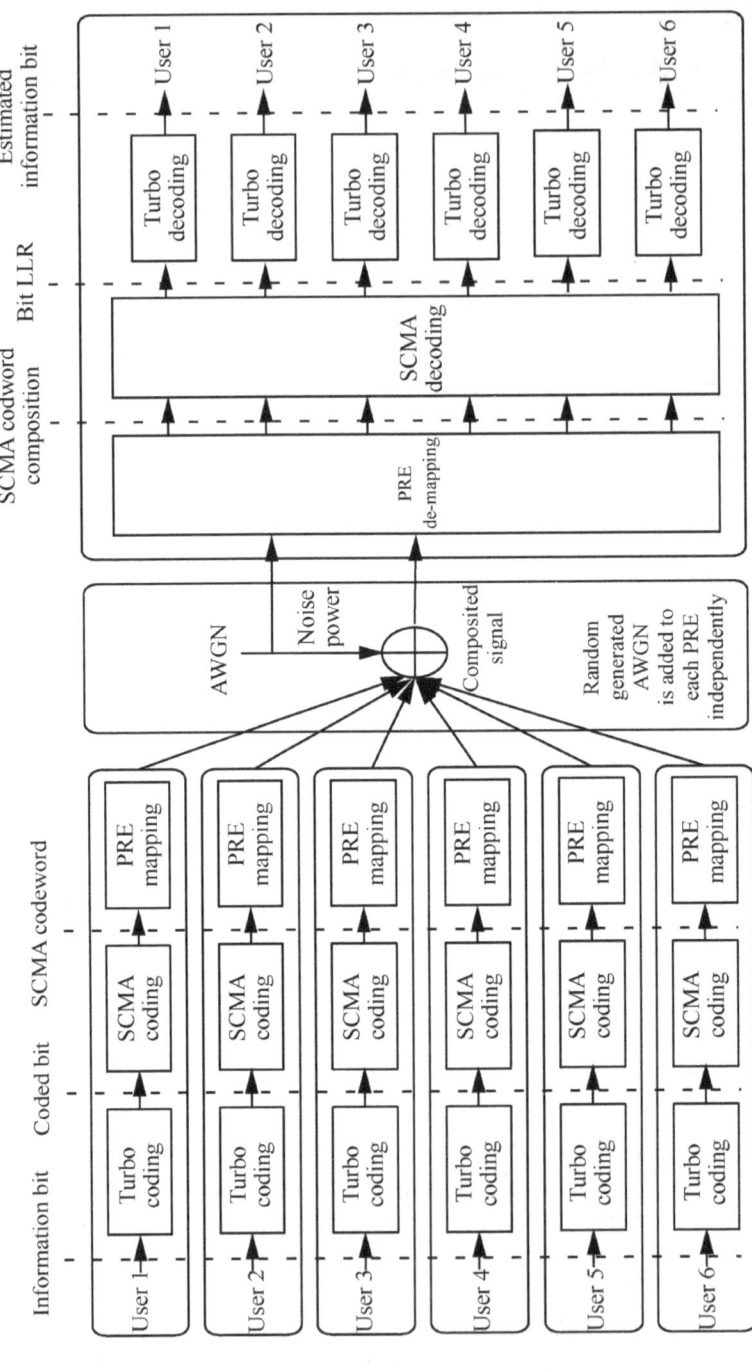

Figure 3.1: Overall block diagram of uplink SCMA system.

to be considered; Second, the concept of a data layer is introduced in the design of SCMA, which means one-dimension degree of freedom is increased compared to 4G commercial systems, and optimal schemes for multi-user scheduling, users grouping and power allocation need to be studied. Besides, the combination of SCMA and MIMO needs to be further researched, including sparse spreading in space domain and so on.

3.1.2 Transmission Scheme and Key Technology

1. Transmitter Schemes and Key Technology

The procedures of an SCMA transmitter are illustrated in Figure 3.2 [4, 5]. There are multiple data layers in SCMA for multi-user multiplexing. Data from a single user can be divided into one or several layers. Each data layer has a pre-defined codebook. Each codebook contains several codewords composed of multidimensional symbols. Codewords belong to the same codebook have the same sparse pattern. For each data layer, the bits after channel encoding are mapped to codewords directly. On account of SCMA, data from different users can be multiplexed in the code domain and power domain and share same time-frequency resources.

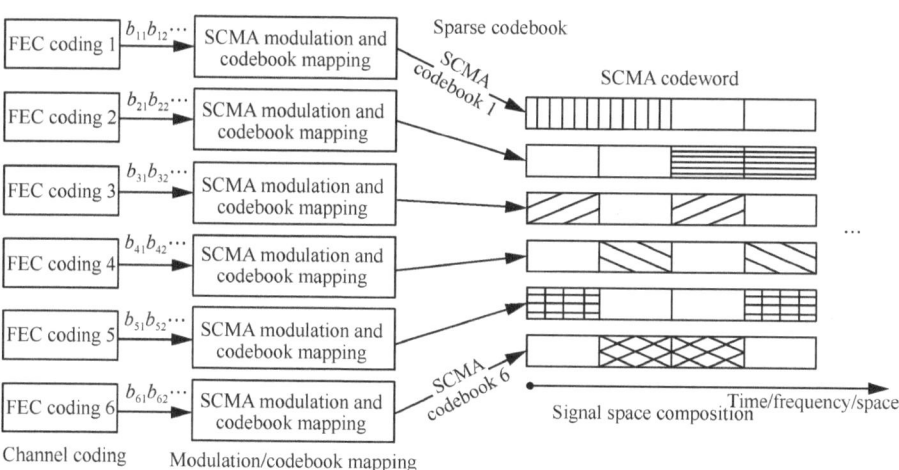

Figure 3.2: Procedures of SCMA transmitter.

Compared with orthogonal multiple access (OMA), NOMA can contain more data and improve the overall system capacity, while the detection complexity at the receiver is increased at the same time. The receiver detection complexity of SCMA is affordable. It is benefited from two following features of SCMA: one is the sparsity of SCMA codewords; the other is the constellation reduction technology when designing the SCMA codebook [1], which is based on multidimensional modulation constellation point order reduction and projection.

The codebook design is a key factor for ensuring the SCMA system performance and flexibility. SCMA codebook design can be viewed as a joint optimization of multidimensional modulation and LDS. Compared to simple repetition coding, it provides more coding gain to SCMA. Reference [1] gives a specific description of SCMA codebook design. In the article, a multidimensional constellation with good Euclidean Distance is designed, which is called basic constellation; based on the basic constellation and by different operations (such as phase rotation), multiple sparse codebooks with different layers are constructed.

Figure 3.3 describes the principle of modulation and mapping with reduced order and projective codebooks based on constellation points reduction [1, 3].

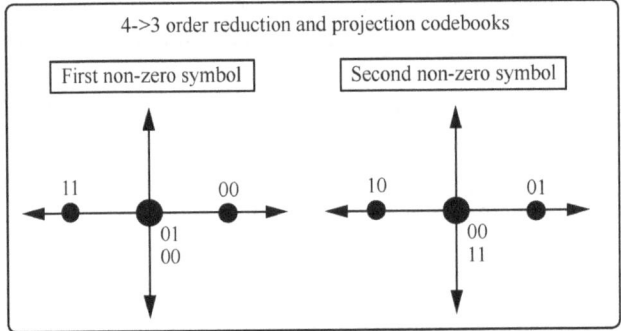

Figure 3.3: SCMA codebook design based on order reduction and projection.

SCMA applies a similar layer mapping method with LTE. One user/data stream can be allocated to one or multiple SCMA data layers. The difference is the mapping from information bits to codewords needs to be realized in each SCMA data layer. SCMA is a sparse code, which means that most elements in the codeword sequence is zero. For the codewords belonging to the same SCMA data layer, positions of the zero elements are the same. The mapping principle [3] is illustrated in Figure 3.4. Supposing there are six users in the SCMA system, which means six data layers and six corresponding codebooks. Each codebook contains 4 codewords whose length is equal to 4. In the process of mapping, the codeword is chosen from the codebook according to the bit number, and codewords from different data layers are superposed directly.

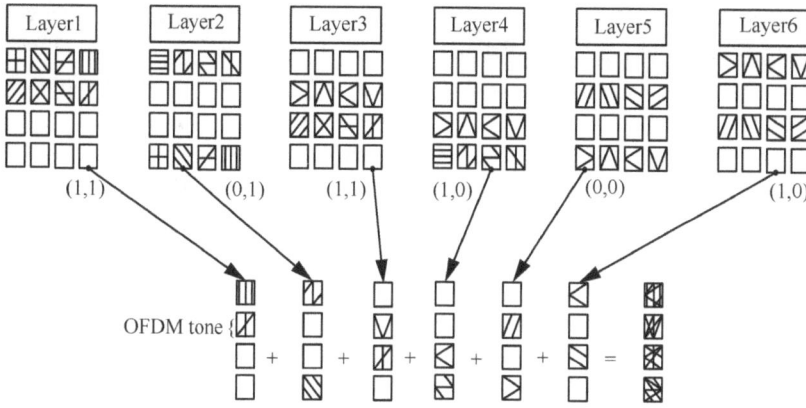

Figure 3.4: Illustration of bit to codeword mapping.

2. Receiver Schemes and Key Technology

Advanced receivers need to be employed at the SCMA receiver to carry on multi-user detection and decoding. In uplink grant-free transmissions, the information of transmitted users is unknown to the base station, thus the receiver needs to do active user detection to the channel. Figure 3.5 is the workflow of detecting active users at the receiver. The receiver first detects the active users and then executes information detection to these active users. Due to the interference and noise in the channel, some active users may be considered as unused, vice versa. These misjudged users are called lost users and false alarm users. The design schemes at the receiver consider the misjudging cases when detecting active users. The SCMA receiver has strict requirements for decreasing complexity and processing delay.

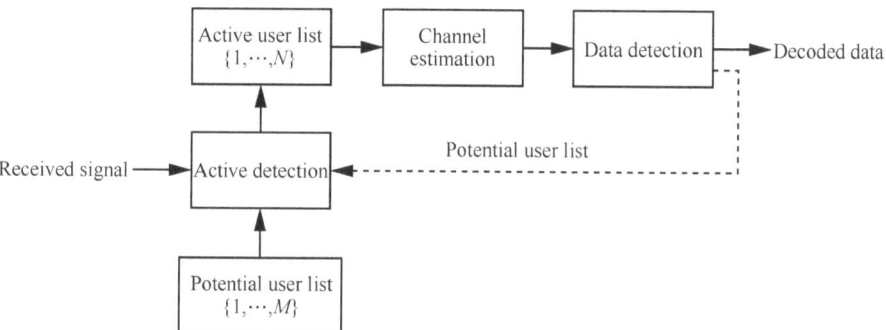

Figure 3.5: Workflow of uplink grant-free receiver.

In the 3GPP RAN WG1 #86b meeting, Huawei submitted SCMA simulation results of SIC-MPA and Expectation Propagation Algorithm (EPA) receivers [6]. The results indicate that SIC-MPA receiver with random codebook allocation and MPA receiver with fixed codebook allocation have the similar performance; the EPA receiver shows robustness when users collide with each other. Therefore, it is suitable for uplink grant-free transmission.

Moreover, a lot of scholars researched receiver algorithms with lower complexity. Reference [7] proposes Partial Marginalization (PM) detection algorithm, which can achieve the BER performance of the MPA algorithm with a low complexity. Reference [8] suggests to simplify the SCMA detection algorithm by a priori judgement to users with high confidence.

3.1.3 Simulation Evaluation and Performance Analysis

1. Link-Level Simulation and Performance Analysis

(1) Uplink Link-Level Simulation

In the 3GPP RAN WG1 #86b meeting, Huawei provided the uplink SCMA link-level simulation results, with simulation parameters listed in Table 3.1.

Table 3.1: Link-level simulation parameters.

Parameter	Value
Carrier frequency	2 GHz
Waveform	OFDM
Carrier parameters	Same as Release 13
System bandwidth	10 MHz
Total transmission bandwidth	4 RB
Target spectrum efficiency	Spectrum efficiency per user: $[0.01\ 0.5]$bit/s $\cdot$ Hz^{-1}
Channel coding	LTE Turbo
BS antenna number	2/4 Rx
UE antenna number	1 Tx
Transmission mode	TM1
Distribution of received average SNR	Equal/Unequal
Multiplexed user number	4, 6, 8, 12

Table 3.1 (continued)

Parameter	Value
Channel model and UE moving speed	TDL-A 30 ns, 3 km/h
Maximum Hybrid Automatic Repeat reQuest (HARQ) transmission number	1
Channel estimation	Actual/Ideal

With the same average received SNR and fixed codebook allocation, the throughput performance gain of SCMA relative to OFDMA is given in Table 3.2. From the table, it can be seen that, when the spectrum efficiency of SCMA for users is above 0.05 bit/s $\cdot$ Hz^{-1} and the number of users is larger than 4, SCMA can obtain a higher throughput than OFDMA. Reference [9] presents the throughput performance comparison among SCMA, OFDMA, LDS-OFDMA and MC-CDMA in actual channel estimation situations. Simulation results indicate that the uplink non-orthogonal multiple access scheme can obtain a higher throughput than OFDMA in both ideal and actual channel estimation situations.

Table 3.2: Performance gain of SCMA@TDL-A 30 ns, 1 Tx2 Rx.

	4UE					6UE				
User spectrum efficiency/bit $\cdot$ s$^{-1}\cdot$ Hz^{-1}	0.05	0.1	0.2	0.3	0.4	0.05	0.1	0.2	0.3	0.4
Gain/dB	0.1	0.5	1.8	1.9	2.0	0.1	1.4	2.5	3.3	4.3
	8UE					**12UE**				
User spectrum efficiency/bit $\cdot$ s$^{-1}\cdot$ Hz^{-1}	0.05	0.1	0.2	0.3	0.4	0.05	0.1	0.2	0.3	0.4
Gain/dB	0.6	2.1	3.4	4.7	5.8	1.7	3.4	6.1	9.3	10.5

In fixed or random codebook allocation situations, the BLER curves versus SNR with eight users and different receiving antennas in the SCMA system are illustrated in Figure 3.6. From the figure, it can be seen that SCMA has high robustness when codebooks collide.

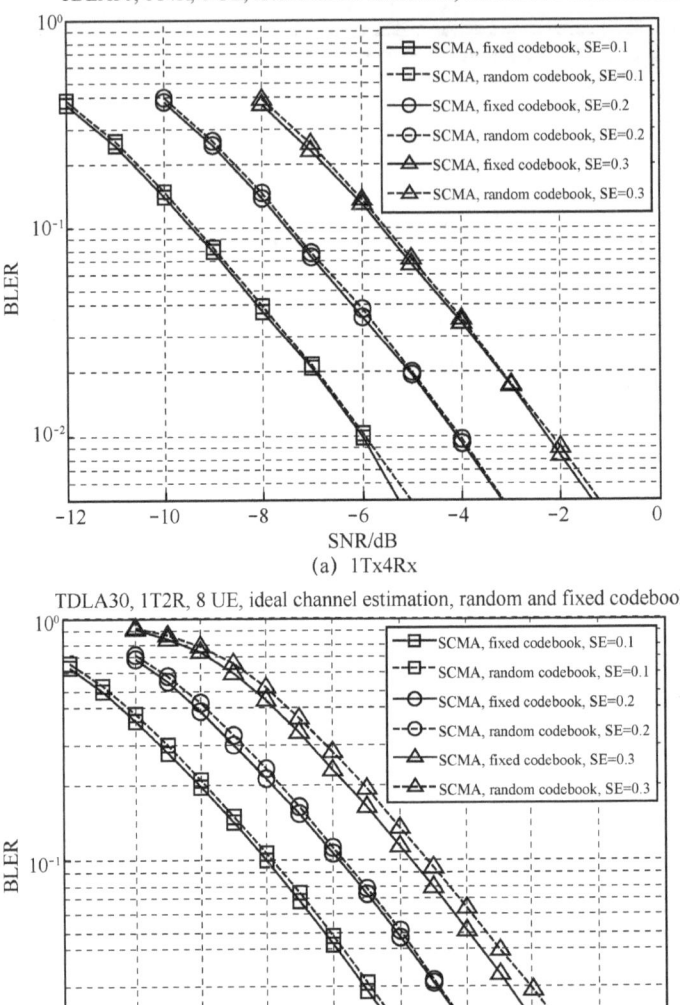

TDLA30, 1T4R, 8 UE, ideal channel estimation, random and fixed codebook

(a) 1Tx4Rx

TDLA30, 1T2R, 8 UE, ideal channel estimation, random and fixed codebook

(b) 1Tx2Rx

Figure 3.6: BLER curves with fixed and random codebook allocation in SCMA.

(2) Downlink Link-Level Simulation

In the 3GPP RAN WG1 #86 meeting, Huawei also showed the downlink SCMA link-level simulation results [10]. When the multiplexed user number is 2, SCMA can obtain a larger capacity region than OFDMA and improve the system throughput in both single user and multi-user situations. Figure 3.7 illustrates the capacity regions of OFDMA and SCMA when the SNR of the two users is [5,20] dB and [10,20] dB, respectively.

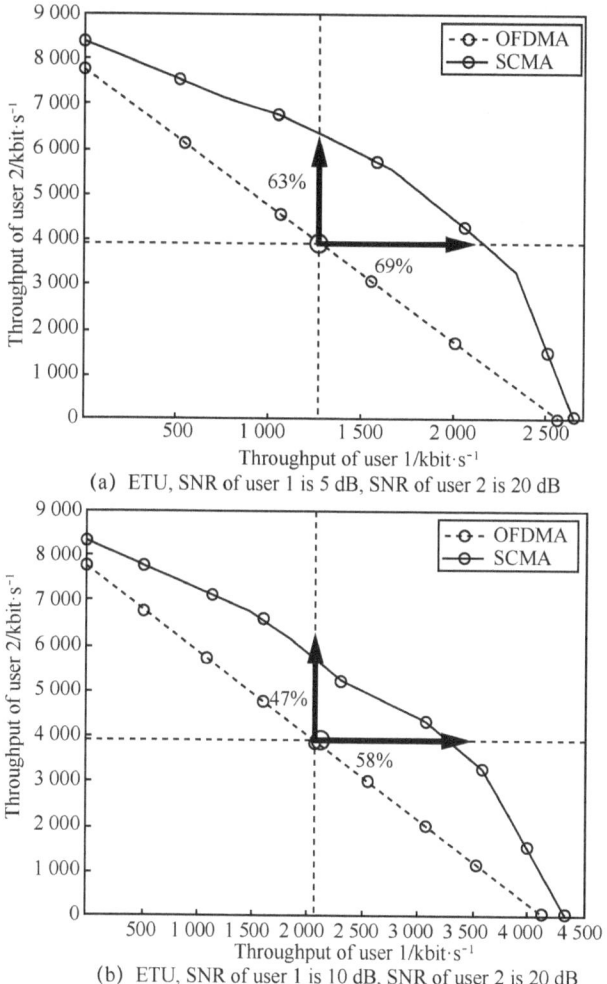

(a) ETU, SNR of user 1 is 5 dB, SNR of user 2 is 20 dB

(b) ETU, SNR of user 1 is 10 dB, SNR of user 2 is 20 dB

Figure 3.7: Capacity region curves of two-user case in Extended Typical Urban (ETU) channel.

From Figure 3.7, it can be seen that when the SNRs of two users are [5,20] dB and the throughput of user 1 is the same; user 2 in SCMA systems can achieve a through-put gain of 63% compared with OFDMA; similarly, while the throughput of user 2 is the same, user 1 in SCMA systems can achieve a throughput gain of 69% compared with OFDMA. When the SNRs of two users are [10,20] dB, while the throughput of user 1 is the same, user 2 in SCMA systems can achieve a throughput gain of 47% compared with OFDMA; similarly, while the throughput of user 2 is the same, user 1 in SCMA systems can achieve a throughput gain of 58% compared with OFDMA. These results indicate that SCMA could obtain a larger capacity region than OFDMA.

2. System Level Simulation and Performance Analysis

(1) Uplink System-Level Simulation

In the 3GPP RAN WG1 #86 meeting, Huawei provided the system-level simulation results of uplink SCMA based on the mMTC scenario [4]. The system performance is evaluated by measuring the system outage probability under different loads. Figure 3.8 describes the system outage probability curves of SCMA and OFDMA versus system load in grant free competitive access systems.

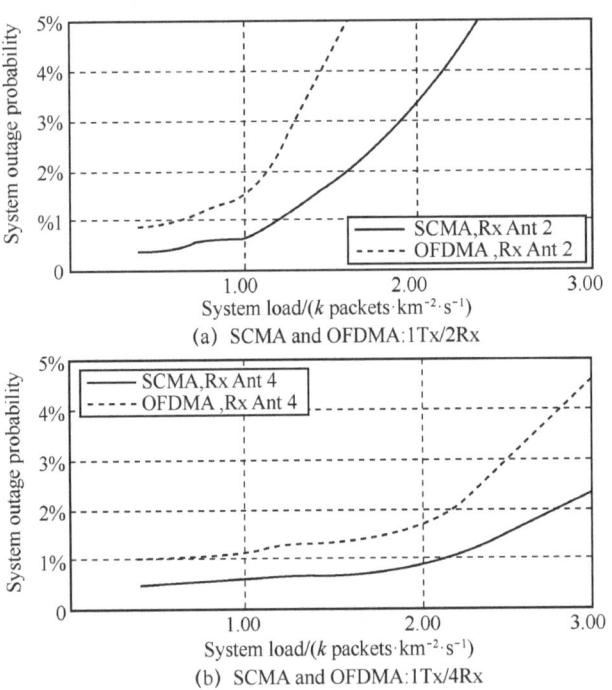

Figure 3.8: System performance comparison between SCMA and OFDMA.

Simulation results indicate that the system outage probability of SCMA can be improved obviously compared with OFDMA. When the system outage probability is 1%, compared with OFDMA baseline, system load of SCMA in Rx and 4 Rx cases improve by 97% and 233%, respectively, which means the capacity of SCMA is 2–3 times of OFDMA.

(2) Downlink System-Level Simulation

In the 3GPP RAN WG1 #86 meeting, Huawei also summarized system-level simulations of downlink SCMA with full buffer and non-full buffer in the eMBB scenario [11]. In full buffer open loop MIMO scenario, the multi-user SCMA could obtain a spectrum efficiency gain of 24%–32% compared with OFDMA. In non-full buffer scenario, if just considered user throughput with small packet, the SCMA can also achieve an obvious gain compared with OFDMA.

3.1.4 Prototype Develop and Test Evaluation

Apart from the simulation verifications, Huawei also carried on prototype implementations and test verifications to SCMA [2, 5]. The prototype system of SCMA is a soft baseband platform based on the concept of software defined radio. In the system, Huawei RH2288 server is used to implement the baseband of the base station, and Huawei RRU 3232 radio frequency (RF) module is used to implement the RF of the base station. The baseband of the UE is implemented by MacBook laptop, which is connected with an external mobile RF module. In lab tests, the base station and the UE are connected to a channel simulator.

To further verify the performance of SCMA, Huawei deployed the UE to different positions by designing different test cases and made field tests with the platform in different situations. The field test scenario is illustrated in Figure 3.9. The prototype system can work in OFDMA or SCMA mode and support real-time changes of different modes. Moreover, new waveform filtered-OFDM (f-OFDM) can be further combined with SCMA to improve spectrum efficiency and decrease out-of-band loss. Test results

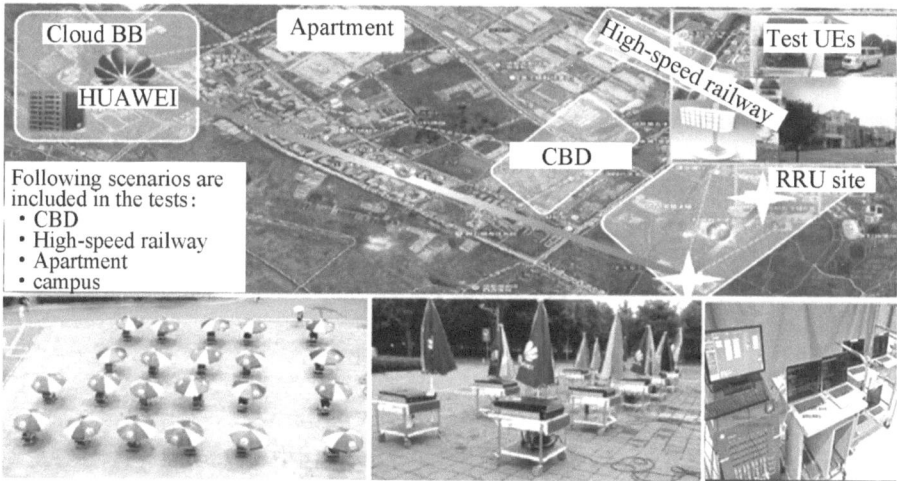

Figure 3.9: Prototype verify of SCMA.

show the combination of SCMA prototype and f-OFDM can not only make better utilization of local spectrum but also support asynchronized access among different SCMA
groups. Reference [5] presents field tests and performance analysis of the SCMA prototype system. Preliminary results indicate that simultaneous access of 12 users can be
supported by SCMA in 4 Resource Blocks (RB) in the uplink, which is a 300% overload
transmission; SCMA can support non-orthogonal multiplexing of 3 users in the downlink, with a throughput improvement of more than 50%.

3.2 PDMA

Bin Ren

3.2.1 Technical Principle

PDMA [2–14] is a NMA technology based on characteristic patterns. The transmitter
maps signals from multiple users to corresponding resources according to the pattern
matrix, and the receiver separates superposed signals by advanced receivers (such as
SIC, BP-IDD and so on). Figures 3.10 and 3.11 show the uplink and downlink example
of PDMA, respectively.

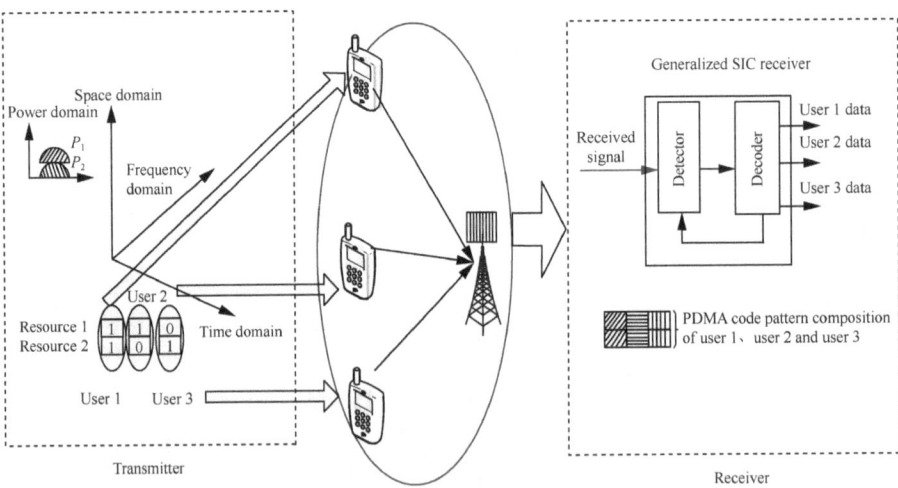

Figure 3.10: The uplink example of PDMA.

In the system illustrated in Figure 3.10, non-orthogonal transmission is carried on
with three users multiplexed on two resource blocks. The resources mentioned here
can be time-frequency resource, frequency domain resource, space domain resource,
power domain resource and their combinations.

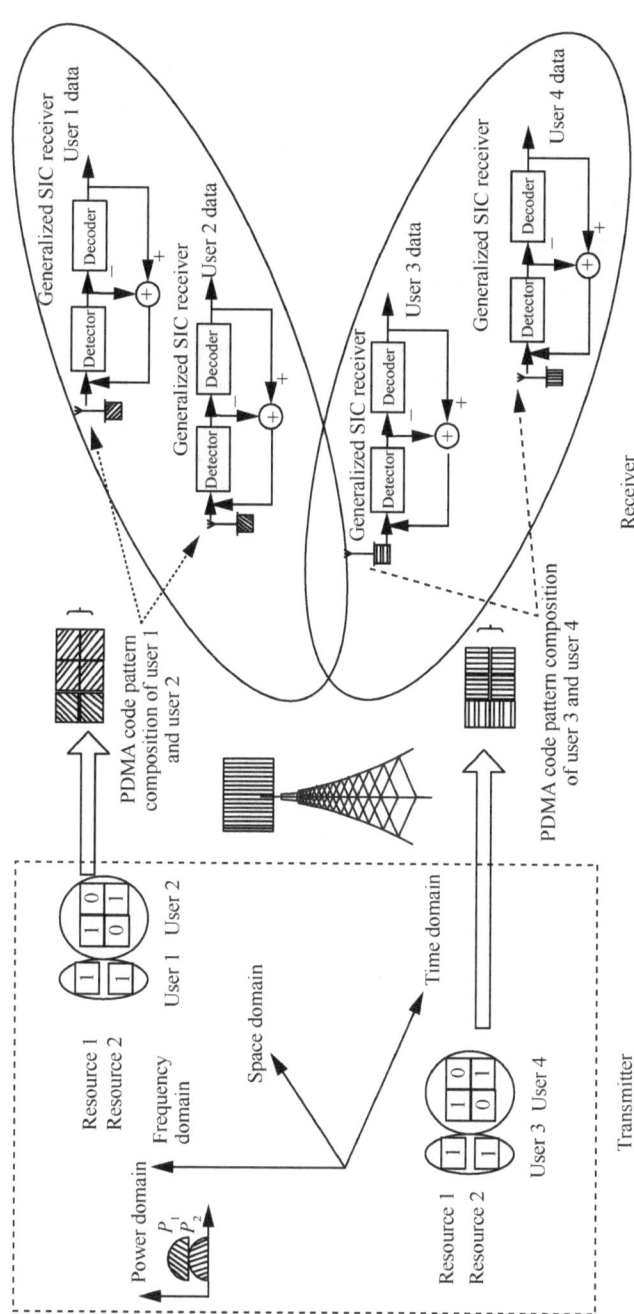

Figure 3.11: The downlink example of PDMA.

In the system illustrated in Figure 3.11, user 1 and user 2 are in one beam direction, while user 3 and user 4 are in another beam direction. Therefore, multi-user MIMO can be used to distinguish users in different beam directions. However, the users in the same beam direction cannot be distinguished by space division multiplexing. In this situation, PDMA characteristic pattern in time-frequency domain can be used to distinguish users and realize non-orthogonal transmission.

Figure 3.12 is the PDMA system model. In this model, single antenna is configured at both the transmitter and the receiver [13].

Figure 3.12(a) is the uplink of the PDMA system model. At the transmitter, PDMA encoder maps modulation symbols to corresponding RBs and generates PDMA modulation vectors. At the receiver, the PDMA recovers data from multiple users with multi-user detectors.

Suppose there are K users in the uplink system, the user data is mapped to N RBs by different characteristic pattern $\mathbf{g}_k$. The PDMA modulation vector $\mathbf{v}_k$ of user k is:

$$\mathbf{v}_k = \mathbf{g}_k x_k, \quad 1 \le k \le K \tag{3.1}$$

where $\mathbf{g}_k$ is a $N \times 1$ binary vector, including elements "1" and "0". Element "1" denotes the user data is mapped to corresponding RB, while element "0" denotes no user data is mapped to corresponding RB. A PDMA pattern matrix with K users multiplexing on N RBs can be expressed by $\mathbf{G}_{PDMA}^{[N,K]}$ with N rows and K columns, where $\mathbf{G}_{PDMA}^{[N,K]} = [\mathbf{g}_1, \mathbf{g}_2, \cdots, \mathbf{g}_K]$.

The received signal at the base station is:

$$\mathbf{y} = \sum_{k=1}^{K} diag(\mathbf{h}_k)\mathbf{v}_k + \mathbf{n} \tag{3.2}$$

where $\mathbf{n}$ is the received noise and interference; $\mathbf{h}_k$ is the uplink channel response of user k; $\mathbf{y}$, $\mathbf{n}$ and $\mathbf{h}_k$ are all vectors with length N; $diag(\mathbf{h}_k)$ denotes a diagonal matrix with the diagonal element equals $\mathbf{h}_k$. Equation (3.2) can be simplified as follows:

$$\mathbf{y} = \mathbf{H}\mathbf{x} + \mathbf{n} \tag{3.3}$$

where $\mathbf{x} = [x_1 \quad x_2 \quad \cdots \quad x_k]^T$; $\mathbf{H}$ is the equivalent channel response matrix with K users multiplexing on N RBs; and $\mathbf{H} = \mathbf{H}_{CH} \cdot \mathbf{G}_{PDMA}^{[N,K]}$, in which $\mathbf{H}_{CH} = [h_1 \quad h_2 \quad \cdots \quad h_k]$ is the real channel response matrix between the K users and the base station. "$\cdot$" denotes multiplication of elements in corresponding positions of the two matrixes.

When $\mathbf{G}_{PDMA}^{[3,6]} = \begin{bmatrix} 1 & 1 & 0 & 1 & 0 & 0 \\ 1 & 0 & 1 & 0 & 1 & 0 \\ 0 & 1 & 1 & 0 & 0 & 1 \end{bmatrix}$, the received signal of the base station can be expressed as:

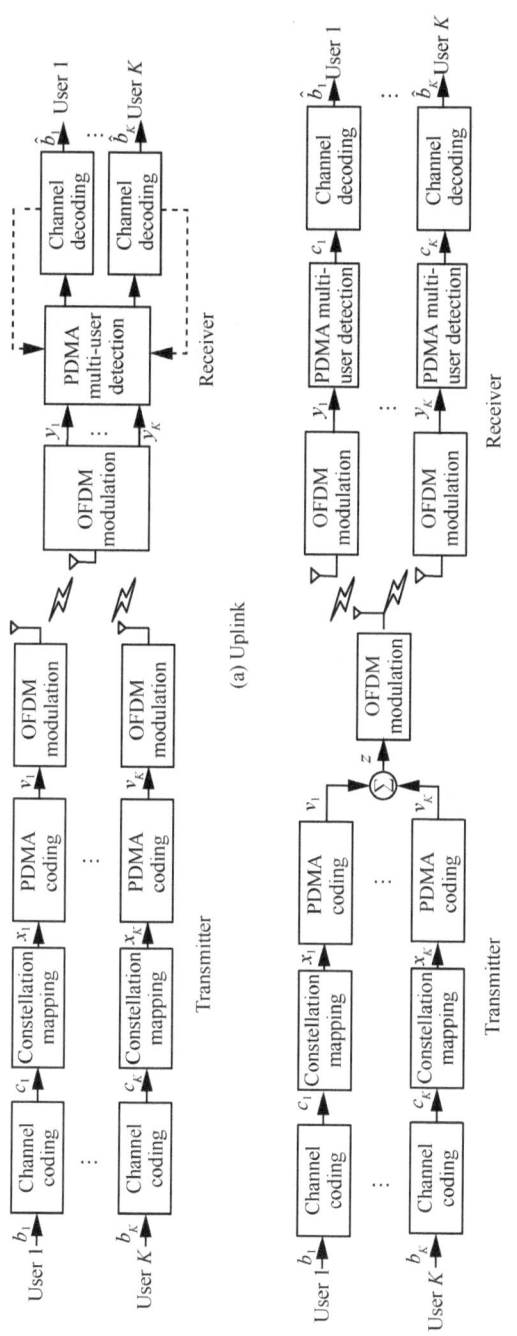

Figure 3.12: PDMA system model.

$$
\begin{bmatrix} y_1 \\ y_2 \\ y_3 \end{bmatrix} = \begin{bmatrix} h_{1,1} & h_{1,2} & 0 & h_{1,4} & 0 & 0 \\ h_{2,1} & 0 & h_{2,3} & 0 & h_{2,5} & 0 \\ 0 & h_{3,2} & h_{3,3} & 0 & 0 & h_{3,6} \end{bmatrix} \begin{bmatrix} x_1 \\ x_2 \\ x_3 \\ x_4 \\ x_5 \\ x_6 \end{bmatrix} + \begin{bmatrix} n_1 \\ n_2 \\ n_3 \end{bmatrix} \tag{3.4}
$$

Figure 3.12(b) is the downlink of the PDMA system model. Suppose each user is allocated a characteristic pattern. After PDMA coding, multiple data streams are superposed at the base station and transmitted. The received signal of user k can be expressed as:

$$
\mathbf{y}_k = diag(\mathbf{h}_k) \sum_{i=1}^{K} g_i x_i + \mathbf{n}_k = \left[diag(\mathbf{h}_k) \mathbf{G}_{PDMA}^{[N,K]} \right] \mathbf{x} + \mathbf{n}_k = \mathbf{H}_k \mathbf{x} + \mathbf{n}_k \tag{3.5}
$$

where $\mathbf{H}_k = diag(\mathbf{h}_k)\mathbf{G}_{PDMA}^{[N,K]}$; $\mathbf{n}_k$ is the noise and interference at the receiver; $\mathbf{h}_k$ is the downlink channel response of user k; $\mathbf{H}_k$ is the equivalent channel response matrix of user k; $\mathbf{x} = [x_1 \quad x_2 \quad \cdots \quad x_k]^T$; x_k is the modulation symbol of user k.

3.2.2 Transmission Scheme and Key Technology

1. PDMA Mapping Scheme
In PDMA systems, the characteristic pattern defines the mapping way of each user to resource groups. The characteristic pattern can be expressed by binary vector, whose length is the same as the RB number in the resource group. There is a one-to-one correspondence between each element in the pattern matrix and one RB in the resource group. Figure 3.13 takes 6 users multiplexing on 4 RBs as an example to introduce the resource mapping scheme of PDMA. Each user is allocated a characteristic pattern. The data of user 1 is mapped to 4 RBs, the data of user 2 is mapped to the first 3 RBs and so on. The diversity degree of the 6 users are 4, 3, 2, 2, 1 and 1 respectively [2, 13].

$$
\mathbf{G}_{PDMA}^{[4,6]} = \begin{bmatrix} 1 & 1 & 1 & 0 & 0 & 0 \\ 1 & 1 & 0 & 1 & 0 & 0 \\ 1 & 1 & 1 & 0 & 1 & 0 \\ 1 & 0 & 0 & 1 & 0 & 1 \end{bmatrix} \tag{3.6}
$$

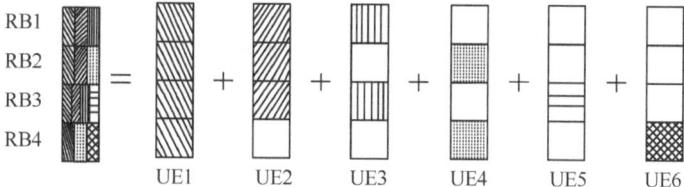

Figure 3.13: PDMA resource mapping.

2. PDMA Pattern Design

Each column of the pattern matrix represents a characteristic pattern in PDMA, and the properties of the matrix, such as the dimension and the sparsity, affects the system performance and the receiver complexity directly. If K and N are chosen, the overload rate ($\alpha = K/N$) is determined. For example, a system with an overload rate equals 150% can be implemented through a pattern matrix with 2 rows and 3 columns. It means the data of 3 users multiplexed on 2 RBs. The corresponding pattern matrix is:

$$\mathbf{G}_{PDMA}^{[2,3]} = \begin{bmatrix} 1 & 1 & 0 \\ 1 & 0 & 1 \end{bmatrix} \tag{3.7}$$

Pattern matrix with an overload rate equals 150% can also be represented by a matrix with 4 rows and 6 columns, which is:

$$\mathbf{G}_{PDMA}^{[4,6]} = \begin{bmatrix} 1 & 0 & 1 & 1 & 1 & 0 \\ 1 & 1 & 0 & 1 & 0 & 1 \\ 1 & 1 & 1 & 0 & 1 & 0 \\ 0 & 1 & 1 & 0 & 0 & 1 \end{bmatrix} \tag{3.8}$$

Although these two pattern matrixes have the same overload rate, the performance of $\mathbf{G}_{PDMA}^{[4,6]}$ is obviously better than $\mathbf{G}_{PDMA}^{[2,3]}$. However, the former has a higher detection complexity than the latter, which means its performance gain is obtained at the cost of increasing detection complexity. Pattern matrix with high dimension can achieve a better system performance, but the detection complexity increases at the same time. Thus, with a fixed overload rate, the selection of the pattern matrix needs to consider the balance between the detection complexity and the system performance.

Suppose there are N RBs (the row number of the pattern matrix), $2^N - 1$ different characteristic patterns can then be provided for users to choose. Suppose the user number is K (the column number of the pattern matrix), which can be obtained through calculating with the overload rate when N is given. A PDMA pattern matrix can be got by choosing K characteristic patterns randomly from the $2^N - 1$ characteristic patterns. The PDMA pattern matrix design is important to the system performance and the complexity. The design criterion is as follows [13]:

(1) Choose the corresponding PDMA characteristic pattern matrix dimension according to the system service requirements and the overload rate, then choose suitable column weight based on the computing capability the system can support. If complicated computing can be supported by the system, great column weight (the number of "1" in the pattern) patterns are tend to be chosen, otherwise light column weight patterns are chosen. Greater column weight patterns have higher diversity degree, thus can provide more reliable data transmission; however, the detection complexity at the receiver increases.

(2) The group number (PDMA patterns with same diversity degree are in the same group) with different diversity degrees (the number of "1" in the column in PDMA matrix) in PDMA pattern matrix should be as many as possible, in order to relieve the error propagation impact of SIC receiver or accelerate the convergence speed of BP detection.

(3) As long as the overload rate is large than 100%, there will be patterns with same diversity degree in the PDMA pattern matrix inevitably. In order to optimize the interference cancellation performance as far as possible, the interference among patterns with same diversity degree should be as little as possible. For a given diversity degree, the pattern chosen criterion is minimizing the maximum inner product of any two patterns. For patterns with same diversity degree, the smaller the inner product among different patterns is, the less the interference between each other is. Small inner product indicates there is little overlap between the positions of element "1" of the two patterns, which means shared resource number between the two patterns is little, and data of the two users are only multiplexed on few resources.

The design of the PDMA pattern matrix should consider overload rate, diversity degree and detection complexity. A good pattern matrix can achieve a good balance among these aspects.

In Eq. (3.7), data of 3 users is mapped to 2 RBs, the base station transmitted signal on the 2 RBs is:

$$
\begin{bmatrix} v_1 \\ v_2 \end{bmatrix} = \begin{bmatrix} 1 & 1 & 0 \\ 1 & 0 & 1 \end{bmatrix} \begin{bmatrix} x_1 \\ x_2 \\ x_3 \end{bmatrix}
\tag{3.9}
$$

where v_j denotes the transmitted signal on the jth resource, x_k is the modulation symbol of user k. Different from OMA systems, the signal on PDMA resource is a linear superposition of multiple modulation symbols, which is:

$$
\begin{cases} v_1 = x_1 + x_2 \\ v_2 = x_1 + x_3 \end{cases}
\tag{3.10}
$$

The linear combination may change the characteristics of the transmitted signal. For example, if all the 3 users applies Binary Phase Shift Keying (BPSK) modulation, the modulation symbol of user 1, user 2 and user 3 may be +1 or −1, then the candidate value set of different linear combinations is $\{-2, 0, +2\}$. When there is no noise in the channel, and the received signal on one RB is −2 or +2, the modulation symbols of the two users are $[-1, -1]$ and $[+1, +1]$, respectively. If the received signal on one RB is 0, then there are two possibilities for the modulation symbols of the two users on this RB: $[-1, +1]$ or $[+1, -1]$. In this situation it is impossible for the receiver to make a distinction between the actual transmitted signals of the two users. Furthermore, if modulation scheme such as Quadrature Phase Shift Keying (QPSK) or 16QAM is used, there will be 9 or 49 possibilities for the linear combinations, which make the situation more complicated.

From the above discussion, it can be observed that the constellations after linear combination is non-uniformly distributed; meanwhile, there is no one-to-one mapping between the constellation and the input user data. These will cause "indistinctness" of mapping. In order to eliminate this "indistinctness," a power scaling factor β and a phase rotation factor φ are users introduced before the superposition of user symbols. Take two users as an example, the result is:

$$v = \sqrt{\beta} x_1 e^{j\varphi} + \sqrt{1-\beta} x_2 \qquad (3.11)$$

The "indistinctness" problem described above can be solved by applying Eq. (3.11) to the system. The phase rotation factor can still increase the channel capacity gain, which is called forming gain. By applying the power scaling factor and the phase rotation factor into the PDMA pattern matrix, the value "1" in the PDMA pattern matrix will be replaced by a complex number. The amplitude and the phase of this complex number reflect the value of the power scaling factor and the phase rotation factor, respectively. For example, considering a pattern matrix $\mathbf{G}_{PDMA}^{[2,3]}$, the extended PDMA pattern matrix can be expressed as:

$$\mathbf{G}_{E-PDMA}^{[2,3]} = \begin{bmatrix} \alpha_{11}e^{-j\varphi_{11}} & \alpha_{21}e^{-j\varphi_{21}} & 0 \\ \alpha_{12}e^{-j\varphi_{12}} & 0 & \alpha_{32}e^{-j\varphi_{32}} \end{bmatrix} \qquad (3.12)$$

where α_{kj} and φ_{kj} denotes the power scaling factor and the phase rotation factor of user k on the jth RB, respectively. The optimal power scaling factor and phase rotation factor depend on the user number and the shape of the input constellation.

3. PDMA Detection Algorithm
Advanced detection algorithm is the key to achieve performance gain in PDMA uplink and downlink [13]. Detection algorithms suitable for PDMA mainly include: SIC, BP and BP-IDD. The main problem in SIC detection is error propagation. If error exists in an earlier detected data packet, then the detection of the data

packets after this data packet will all be wrong. Error propagation can be reduced to a certain extent by controlling the diversity degree of the PDMA pattern. BP (also known as MPA) separates user information mainly through message delivery between channel nodes and user nodes. BP-IDD (also known as Turbo MPA) algorithm is a joint iteration of Turbo decoding and BP detection actually, whose main principle is treating the bit-level LLR obtained by Turbo decoder as feedback [13]; transforming the bit-level LLR into symbol-level LLR through re-encoding in probability domain and treating the symbol-level LLR as a priori information of the BP detector. There are two main processes included in a BP-IDD detector: internal iteration process of BP detector and external iteration process between BP detector and Turbo. The sparsity of PDMA pattern can reduce the implement complexity of BP and BP-IDD efficiently, which makes BP and BP-IDD more suitable for PDMA systems. Furthermore, the convergence speed of BP and BP-IDD can be accelerated by the PDMA pattern design.

Table 3.3: Computing complexity of each modulation symbol.

Algorithm	Multiplication	Addition
SIC	$O(KN^3)$	$O(KN^3)$
BP	$O(d_f N M^{d_f})$	$O(T_{in} d_f N Q_m M^{d_f})$
BP-IDD	$O(d_f N M^{d_f})$	$O(T_{out} T_{in} d_f N Q_m M^{d_f})$

Reference [13] compares SIC, BP and BP-IDD receiving algorithms. Table 3.3 summarizes the computing complexity of the three algorithms. In the table, M denotes the size of the modulation constellation, $Q_m = lb(M)$. T_{in}, T_{out} and d_f denote BP-IDD internal iteration number, BP-IDD external iteration number and maximum row weight of PDMA pattern matrix, respectively. From the table, it can be seen that the addition operation number of the BP-IDD receiver is T_{out} times as the addition operation number of the BP receiver.

Since the processing capacity of BS is much stronger than user terminal, BP-IDD and BP algorithms are more suitable in PDMA uplink. For downlink transmission, if taking overall consideration on complexity and processing capacity, BP and SIC are appropriate for detection at the UE.

3.2.3 Simulation Evaluation and Performance Analysis

1. Link-Level Simulation and Performance Analysis

(1) Uplink Link-Level Simulation

In the 3GPP RAN WG1 #86 meeting, CATT proposed the latest PDMA uplink link-level simulation results [15]. The uplink link-level simulation parameters are listed in Table 3.4.

Table 3.4: Uplink link-level simulation parameters.

Parameter	Value
Carrier frequency	2 GHz
Waveform	OFDM
Channel encoding	LTE Turbo
Carrier parameters	Same as Release 13
System bandwidth	10 MHz
Total transmit bandwidth	4RB, 12RB
Reference signal overhead	2 DM-RS symbols, no SRS, 144 REs in each RB are used to transmit data
Size of user data (TB size of each user/total RE number for data transmission)	Required transmitted bit number of each user is 4 RB: (no CRC): 120 bit 192 bit 12 RB: (no CRC): 408 bit 624 bit
PDMA pattern matrixes to support different overload rates	$100\%: \mathbf{G}_{PDMA}^{[4,4]} = \begin{bmatrix} 1 & 0 & 0 & 1 \\ 0 & 1 & 1 & 0 \\ 1 & 1 & 0 & 1 \\ 1 & 1 & 1 & 0 \end{bmatrix}$ $200\%: \mathbf{G}_{PDMA}^{[4,8]} = \begin{bmatrix} 1 & 0 & 0 & 1 & 1 & 0 & 0 & 0 \\ 0 & 1 & 1 & 0 & 0 & 1 & 0 & 0 \\ 0 & 1 & 0 & 1 & 0 & 0 & 1 & 0 \\ 1 & 0 & 1 & 0 & 0 & 0 & 0 & 1 \end{bmatrix}$ $300\%: \mathbf{G}_{PDMA}^{[4,12]} = \begin{bmatrix} 1 & 0 & 1 & 1 & 1 & 0 & 0 & 0 & 1 & 0 & 0 & 0 \\ 0 & 1 & 1 & 0 & 0 & 1 & 1 & 0 & 0 & 1 & 0 & 0 \\ 1 & 1 & 0 & 1 & 0 & 1 & 0 & 1 & 0 & 0 & 1 & 0 \\ 1 & 1 & 0 & 0 & 1 & 0 & 1 & 1 & 0 & 0 & 0 & 1 \end{bmatrix}$
BS antenna number	2 Rx
UE antenna number	1 Tx
Transmission mode	TM1 (refer to TS36.213)
Pattern allocation	fix

Table 3.4 (continued)

Parameter	Value
Receiver model (iteration number)	BP-IDD (internal iteration: 3, external iteration: 3)
Average receiving SNR distribution	Equal among users
Channel model	TDL-A (30 ns), TDL-C (300 ns)
UE speed	3 km/h
Maximum HARQ transmission number	1
Channel estimation	Ideal channel estimation

In OMA systems, subcarrier allocation methods include continuous allocation and discrete allocation. In this section, OMA system with discrete subcarrier allocation is used as the baseline. The SNR gains of PDMA compared with OMA in different overload rate and different target spectrum efficiency situations are listed in Table 3.5.

Table 3.5: SNR gains of PDMA compared with OMA (dB@BLER = 10%).

Allocated bandwidth	Channel model	Spectrum efficiency of each user	100%	SNR gain 200%	300%
4 PRB	TDL-A 30 ns	0.25	0.2	3.0	4.3
		0.375	0.1	3.3	4.1
12 PRB		0.25	0.3	3.1	4.4
		0.375	0.2	3.4	4.3
4 PRB	TDL-C 300 ns	0.25	0.4	3.0	4.6
		0.375	1.0	3.4	5.1
12 PRB		0.25	0.5	3.4	5.3
		0.375	1.1	4.4	5.6

From Table 3.5, it can be seen that, in different overload rate and different target spectrum efficiency situations, the PDMA gain compared with OMA is different. When the total spectrum efficiency is the same, PDMA can obtain obvious SNR gain compared with OMA. When the user target spectrum efficiency is 0.25bit/s · Hz^{-1}, with the overload rate growing from 100% to 300%, the SNR gain of PDMA compared with OMA grows from 0.2–0.5 dB to 4.3–5.3 dB; When the user target spectrum efficiency is 0.375bit/s · Hz^{-1}, the SNR gain of PDMA compared with OMA grows from 0.1–1.1 dB to 4.1–5.6 dB. Meanwhile, from Table 3.5 it can still be seen that the SNR gain of PDMA compared with OMA increases with the growth of the over-load and the target spectrum efficiency. In addition, due to the frequency diversity gain introduced by discrete subcarrier allocation, PDMA gain compared with OMA increases with the growth of Physical RB (PRB) number.

Besides, reference [15] gives the robustness analysis of PDMA in different overload rate situations. Simulation results are illustrated in Figure 3.14. When the target BLER is 10%, the SNR dynamic amplitude with different overload rates in Figure 3.14(a) is 1.1 dB, and the SNR dynamic amplitude with different overload rates in Figure 3.14(b) is 2.8 dB. The performance of 300% overload rate is quite different from the performance of 100% overload rate and 200% overload rate, while the performance of 100% overload rate is close to the performance of 200% overload rate. Simulation results indicate that PDMA can carry on grant-free transmission robustly with different overload rates.

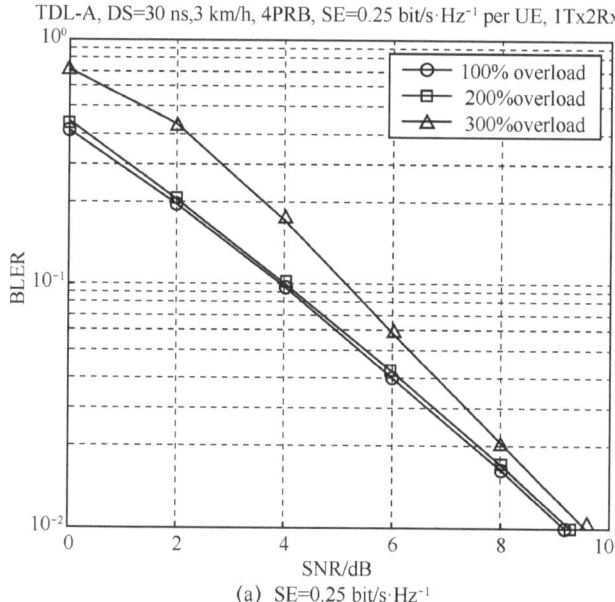

TDL-A, DS=30 ns,3 km/h, 4PRB, SE=0.25 bit/s·Hz^{-1} per UE, 1Tx2Rx

(a) SE=0.25 bit/s·Hz^{-1}

Figure 3.14: BLER curves of PDMA with different overload rates.

TDL-A, DS=30 ns, 3 km/h, 4PRB, SE=0.375 bit/s·Hz^{-1} per UE, 1Tx2Rx

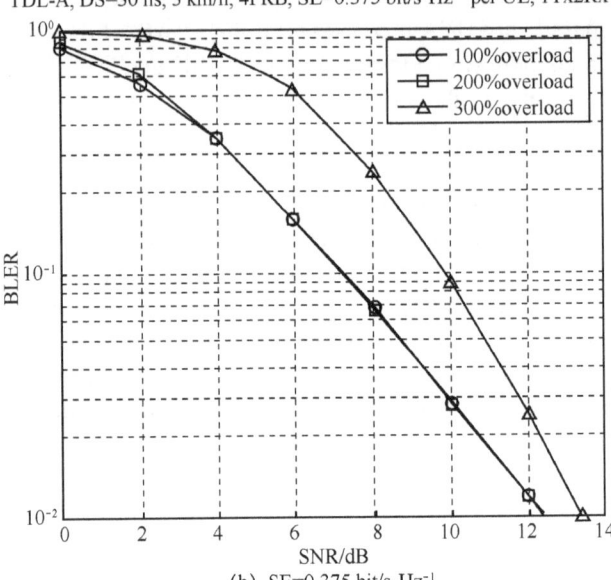

(b) SE=0.375 bit/s·Hz^{-1}

Figure 3.14 (continued)

(2) Downlink Link-Level Simulation

In reference [16], CATT presents the PDMA downlink link-level simulation results. The simulation parameters are listed in Table 3.6. Figure 3.15 shows the throughput comparison between UE far away from the BS and UE near the BS in PDMA scheme with 4 RB, ETU channel and 3 km/h UE speed. The power allocation proportions of UE far away from the BS and UE near the BS are {0.6, 0.4}, {0.7, 0.3}, {0.8, 0.2} and {0.9, 0.1}, respectively. MCS corresponding to QPSK is chosen dynamically, and the code rate set is {0.1, 0.2, 0.3, 0.4, 0.5, 0.6, 0.7, 0.8}. From Figure 3.15, it can be seen that compared with OMA, UE near the base station and UE far away from the base station in PDMA systems can both achieve a higher throughput, and the achieved throughput is different with different power allocation proportions. When the power allocation proportion is {0.7, 0.3}, the sum throughput is the highest. It is thus clear that PDMA can obtain a higher system throughput by appropriate power allocation.

Table 3.6: Downlink link-level simulation parameters.

Parameter	Value
Carrier frequency	2 GHz
Waveform Channel encoding	OFDM LTE Turbo
Carrier parameters	Same as Release 13
System bandwidth	10 MHz
Transmit bandwidth allocation	4RB (0.72 MHz)
Reference signal and control signaling overhead	2 PDCCH symbols, one CRS port in TM1, 2 138 REs in each RB are used to transmit data
Target spectrum efficiency	MCS: dynamic, QPSK, code rate set is {0.1, 0.2, 0.3, 0.4, 0.5, 0.6, 0.7, 0.8}. Overload: 2 users, PDMA pattern matrix 4×8. UEs far away from the base station occupy the first 4 rows, UEs near the base station occupy the last 4 rows.
BS antenna configuration	2 Tx
UE antenna configuration	2 Rx
Transmission mode	SFBC (TM2)
Multiple user SNR allocation	Fixed spacing with {0, 10} dB among UEs
UE number	2
SNR of Reference UE	0 dB
Power allocation among UEs	Dynamic: {0.6, 0.4} {0.7, 0.3} {0.8, 0.2} {0.9, 0.1} (power proportion of UE far away from the base station and UE near the base station)
Receiver model (iteration number)	BP-IDD (internal iteration: 3, external iteration: 3)
Propagation channel	ETU
UE speed	3 km/h
Maximum HARQ transmission number	1
Target BLER	0.1
Channel estimation	Ideal

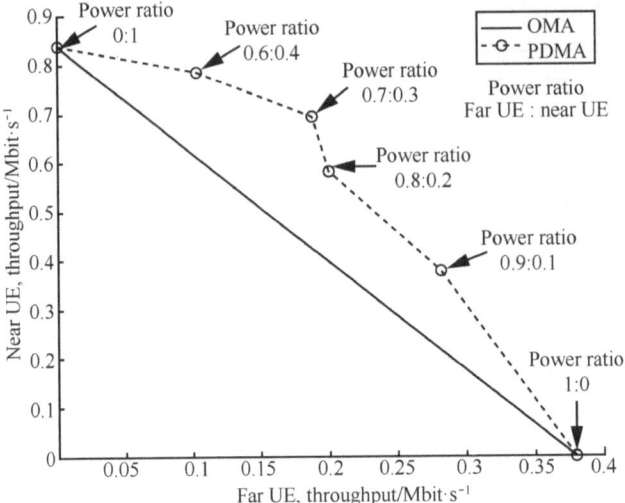

Figure 3.15: Throughput comparison between UE far away from the base station and UE near the base station.

2. System-Level Simulation and Performance Analysis

In the 3GPP RAN WG1 #86 meeting, CATT also submitted system-level simulation results of PDMA system in uplink mMTC scenario [17], and the specific simulation train of thought and method are described in detail in reference [17]. The system-level simulation parameters are listed in Table 3.7.

Table 3.7: PDMA system-level simulation parameters.

Parameter	Value
Layout	Single layer–multiple layer hexagonal grid
Distance between base stations	1,732 m
Carrier frequency	700 MHz
Simulation bandwidth	4 PRB
Channel model	3D Uma
UE transmit power	Maximum 23 dBm
BS antenna number	2 Rx
BS antenna pattern	Same as TR 36.873
BS antenna height	25 m
BS antenna tilt	100°

Table 3.7 (continued)

Parameter	Value
BS antenna gain and link loss	8 dBi, include 3 dB cable loss
Noise coefficient of BS receiver	5 dB
UE antenna number	1 Tx
UE antenna height UE antenna gain	1.5 m −4 dBi
Transmission model	Non full buffer small packet traffic, with packet size equals 20 byte
UE distribution	20% outdoor (3 km/h) 80% indoor (3 km/h)
BS receiver	PDMA: BP-IDD OFMDA: single user MMSE receiver
Uplink power control	Open loop, $P_0 = -95$ dBm and alpha $= 1$
Channel estimation	Ideal channel estimation

Grant free is applied in the simulation. Collision occurs when different users in the PDMA system occupy the same resource group and the same pattern. For OFDMA systems, if users collide with each other, the colliding user signal is considered as interference. However, in PDMA systems, BS will try to detect collided users with BP-IDD algorithm. Simulation results are illustrated in Figure 3.16. It can be seen that, when the packet loss rate equals 1%, the packet arrival rate of OFDMA is 300

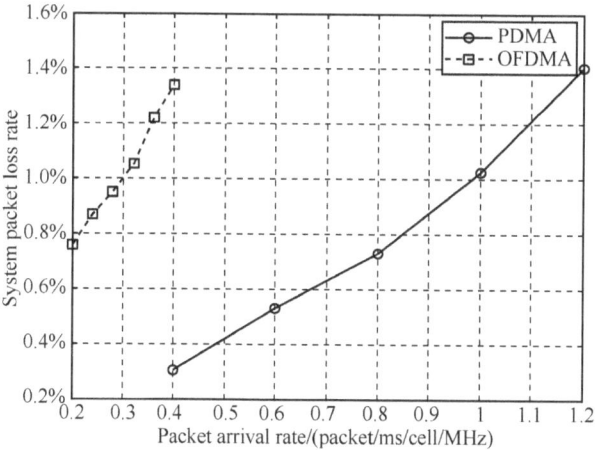

Figure 3.16: System performance comparison between PDMA and OFDMA.

packet/s/cell/MHz approximately, while the packet arrival rate of PDMA is 980 packet/s/cell/MHz approximately. From this we can get that the packet arrival rate of PDMA is 3.26 times the packet arrival rate of OFDMA. In uplink mMTC scenario, PDMA can support more connect users compared with OFDMA.

3.2.4 Prototype Develop and Test Evaluation

The development of PDMA prototype system involves development of the BS and the UE. The BS and the UE are both implemented with Advanced Telecommunications Computing Architecture (ATCA) case and support 20 MHz bandwidth [18]. Take 4 PRB as an example, CATT set up an outdoor verification environment, as illustrated in Figure 3.17, and made comparison tests between PDMA and LTE orthogonal scheme.

Figure 3.17: PDMA test environment.

Consider the uplink system, the UE needs to implement the transmission of uplink Physical Uplink Shared Channel (PUSCH) channel, which includes Turbo coding, scrambling, modulation, physical resource mapping and so on. The BS needs to implement the reception of uplink PUSCH channel, including FFT, channel estimation, BP detection, Turbo decoding and so on. The signal processing flow of the BS and the UE are illustrated in Figure 3.18.

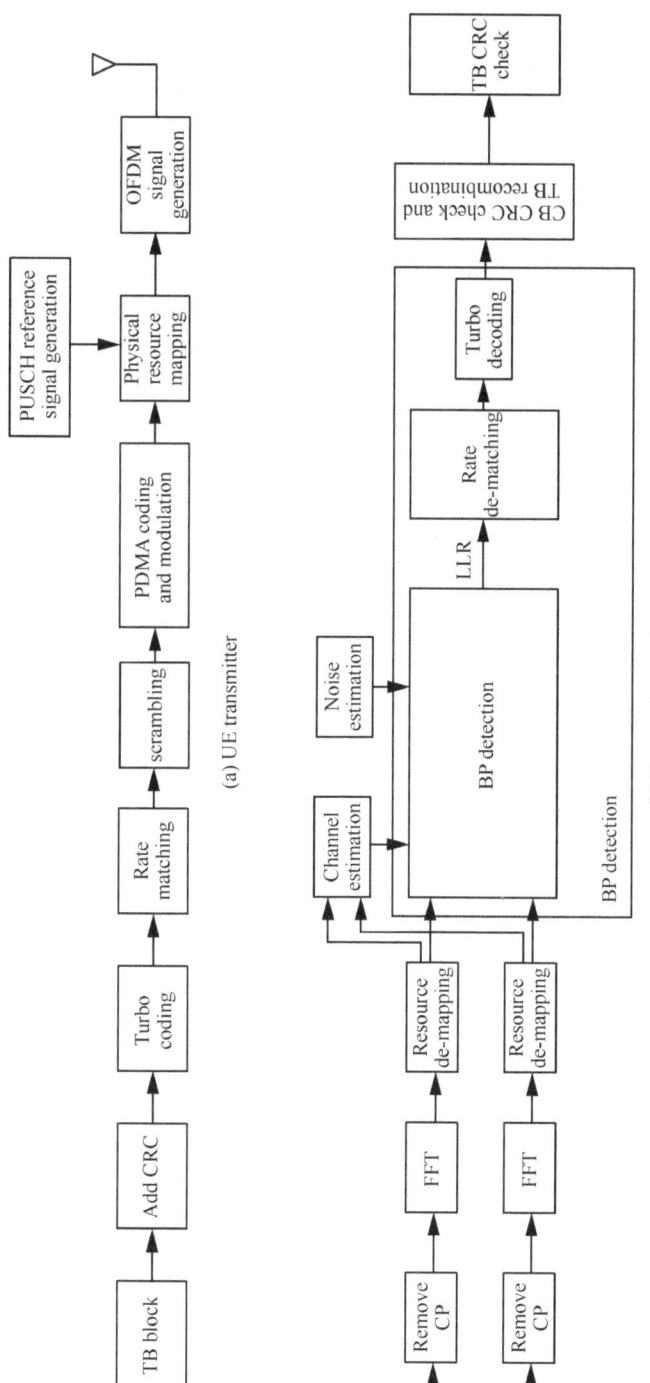

Figure 3.18: Transmit receive processing flow of the PDMA uplink system implementation.

The test results of multiple user connection capability are with the configured 4 PRB, LTE can only access 4 users, while PDMA can access 12 users.

The test results of uplink capacity are as follows: with the configured 4 PRB, after successful access of 4 users the achievable total throughput of LTE is 108.8kbit/s, while after successful access of 12 users, the achievable total throughput of PDMA is 326.4kbit/s.

The test results above show that uplink access user number of PDMA is thrice the number of LTE, the achievable uplink throughput of PDMA after successful access of multiple users is thrice the throughput of LTE.

3.3 SCMA

Xin Su

3.3.1 Technical Principle

MUSA is a non-orthogonal multiple access technology based on complex field poly-cell code [19]. By innovative designed complex field poly-cell code and advanced multi-user detection based on SIC, MUSA can access super large number of users. Figure 3.19 introduces the general uplink procedure of MUSA by considering the case of four resources shared by multiple users [20]. First, different complex spreading sequences are used at the transmitter to spread the symbols of each user. The spreading sequences used here are low cross-correlation sequences and are not necessarily binary

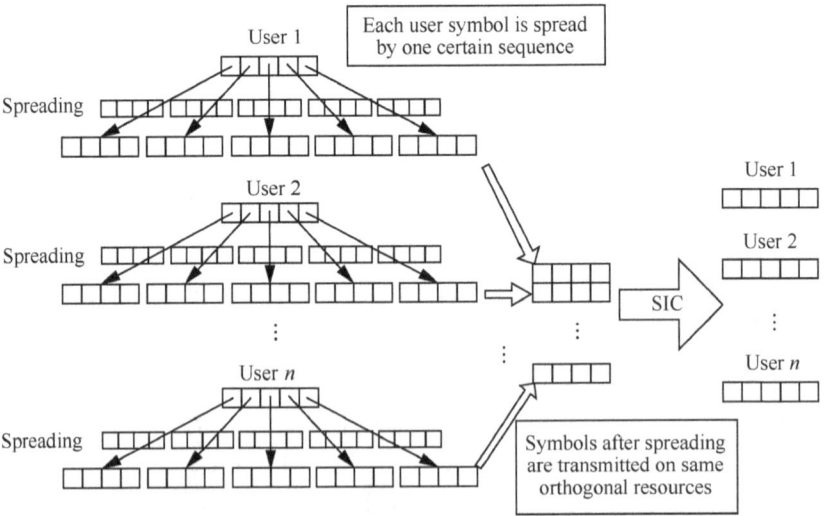

Figure 3.19: Uplink working principle of MUSA.

sequences. Then, symbol sequences after spreading superpose on same time-frequency resources and transmit. CW-SIC is applied at the receiver to separate multi-user signals.

3.3.2 Transmission Scheme and Key Technology

The characteristic of spreading sequences used at the MUSA transmitter affects the system performance and receiver complexity directly. Traditional Direct Sequence-Code Division Multiple Access (DS-CDMA) (such as IS-95 standard) applies long Pseudo-Noise (PN) sequences to ensure low cross-correlation among different sequences, and provide a soft capacity for the system, which means the number of simultaneous access users is larger than the length of the sequences, therefore, the system can work in an overload mode [2]. However, the requirement of massive connections in 5G demands high system overload rate. When the overload rate is high, long PN sequences will make SIC at the receiver rather complicated and inefficient. Thus, MUSA employs complex field poly-cell code as the spreading sequence at the transmitter. When the sequence length is short (such as 8, or even 4), complex sequences can guarantee low correlation. The real part and the imaginary part of each complex number in the complex sequences applied by MUSA belong to a M-ary real number set, such as a simple 3-ary real number set $\{-1, 0, 1\}$, whose constellation is illustrated in Figure 3.20 [21].

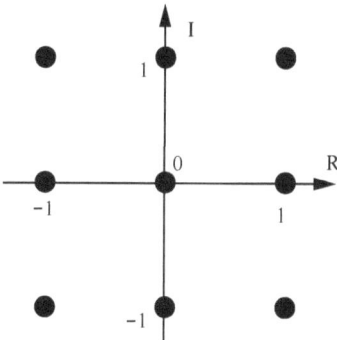

Figure 3.20: Constellation of 3-ary complex sequence element.

MUSA is a grant-free access scheme [22]. UE can access the system autonomously without the schedule of the BS, and each user chooses their own spreading code randomly. For the BS, access user number, spreading code of each user and user channel information are all unknown. Therefore, blind detection is usually applied at the MUSA receiver. Since the complex spreading sequences used in MUSA have short length and limited elements, massive local spreading sequences with low correlation can be generated at the BS. Based on these local spreading sequences and received signal, the performance of blind detection can approach the performance of MMSE detector. There are three parts in the MUSA blind detection: SIC receiver, blind estimation and blind channel estimation. At the receiver, SIC is used by MUSA to filter interference and recover the information of each user.

3.3.3 Simulation Evaluation and Performance Analysis

1. Link-Level Simulation and Performance Analysis

Blind detection is used in the grant-free uplink link-level performance simulation and analysis of MUSA [23]. Simulation parameters are listed in Table 3.8.

Table 3.8: Uplink link-level simulation parameters of MUSA.

Parameter	Value
Carrier frequency	2 GHz
Waveform	OFDM
System bandwidth	10 MHz
Total transmit bandwidth	180 KHz (1 PRB), TTI = 4 ms
Transmission mode	TM1
Multi-user SNR distribution	① [6, 20]dB uniform distribution ② equivalent mean
Length of the spreading sequence	4
Antenna configuration	1 Tx, 2 Rx
UE SNR distribution	Uniform distribution of multiple UE
Channel model and UE speed	TDL-C, 300 ns, 3 km/h
Target BLER	0.1–0.001
UE number	4–30
Modulation Spreading code choice	BPSK random
Information bit of each user	84 (include 24 CRC)
Channel estimation	Actual channel estimation
Channel coding	Turbo 1/2
HARQ	no
Receiver algorithm	Blind CW-SIC

Figure 3.21 gives the BLER performance curve of MUSA with blind detection receiver in different user load situations. It is observed that the BLER of MUSA is less than 10% when the system overload rate is as high as 700% (the user number is 28). When the user number is 4, the BLER of MUSA is about 0.4%. The growth of the BLER performance becomes slower with the increase of the user number when the user number is less than 24.

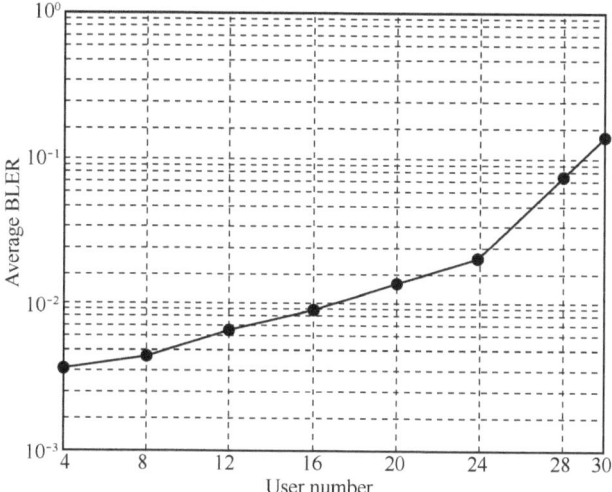

Figure 3.21: BLER performance curve of MUSA with blind detection receiver in different user load situations.

Figure 3.22 gives the BLER versus SNR comparison between MUSA and OFDMA with multi-user equivalent mean SNR. The code rate is 1/2 for both schemes. For OFDMA, supposing there are 4 users with fixed MCS equals to BPSK 1/2. Since link self-adaption needs extra feedback information, which is unrealistic in mMTC scenario. It is easy to see that MUSA has a similar performance with OFDMA when the load is 100% (4 users). When the overload rate reaches 500% (20 users), the performance of MUSA degrades due to the large interference among users. When the overload rate is 400% (16 users), although there is not too much difference in BLER performance between OFDMA and MUSA, MUSA achieves a higher spectrum efficiency than OFDMA.

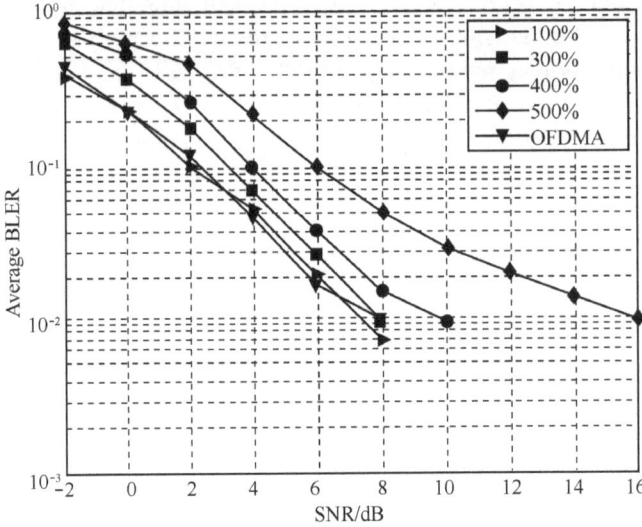

Figure 3.22: SNR and BLER performance curves of MUSA and OFDMA (CR = 1/2).

2. System-Level Simulation and Performance Analysis

For multi-user network model and small data packet traffic model, reference [20] gives the system-level simulation results of MUSA in mMTC scenario. In the 3GPP RAN WG1 #86 meeting, ZTE concluded and explained the simulation assumption and parameters [24]. The updated simulation parameters are listed in Table 3.9. The simulation compares the performance of grant-free MUSA and grant-free SC-FDMA. For grant-free MUSA, each user data occupies the same 4 PRB and generate complex spreading sequences randomly; while for grant-free SC-FDMA, each user chooses 1 PRB for transmission randomly.

Table 3.9: System level simulation parameters of MUSA in mMTC scenario.

Parameter	Value
Layout	Single layer–multiple layer hexagonal grid
Distance between base stations	1,732 m
Carrier frequency	700 MHz
Channel model	3D UMa
UE transmit power	Maximum 23 dBm or choose 10 dBm
BS antenna configuration	Rx: 2 and 4 port (can choose 8)

Table 3.9 (continued)

Parameter	Value
BS antenna height	25 m
BS antenna gain and link loss	8 dBi, include 3 dB cable loss
Noise coefficient of BS receiver	5 dB
UE antenna number	1 Tx
UE antenna height UE antenna gain	1.5 m −4 dBi
Traffic model	Full buffer small packet traffic
UE allocation	20% users in cars outdoors (100 km/h) or 20% users are pedestrians outdoors(3 km/h); 80% users are indoors (3 km/h)
BS receiver	MMSE-IRC
Channel estimation	Actual channel estimation

Figure 3.23 shows the packet loss rate curves of grant-free MUSA and grant-free SC-FDMA in different load situations when the frequency reuse factor is 1 (FR1) and 3 (FR3). From the simulation results, it can be seen that, grant-free MUSA can

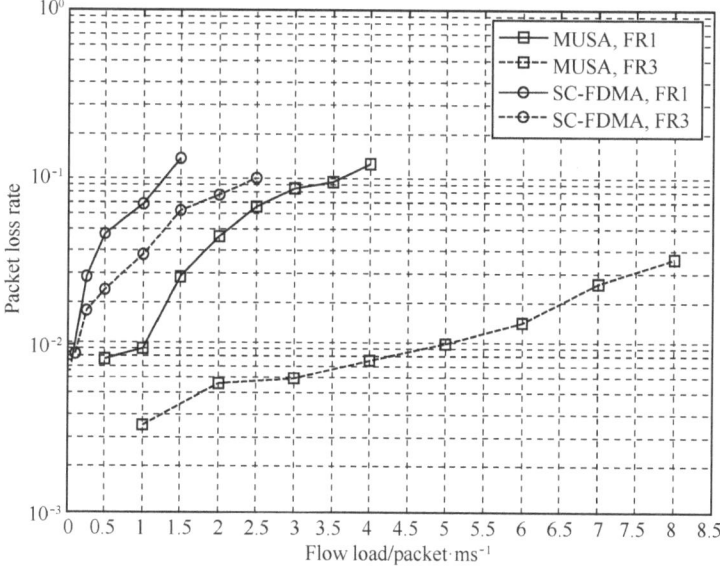

Figure 3.23: Packet loss rate comparison between grant-free MUSA and SC-FDMA.

support a traffic load of 3.5 packet/ms when the frequency reuse factor is 1 and packet loss rate equals 10%. However, the traffic load that grant-free SC-FDMA can support is only 1.25 packet/ms under the same packet loss rate condition. The reason is for grant-free SC-FMDA, when two or more users collide due to selecting the same PRB, the packet loss rate will increase quickly with the growth of the traffic load. It is thus clear that MUSA uses complex spreading sequences with low correlation and is more suitable for grant-free transmissions.

When the frequency reuse factor is 3 and the packet loss rate of grant-free MUSA is 1%, a traffic load of 5 packet/ms can be achieved, which is far better than the supported traffic load (1 packet/ms) when the frequency reuse factor is 1. This is because more serious interference exists in the latter case. However, for SC-FDMA, the gain brought by FR3 compared with FR1 is not as remarkable as MUSA.

3.3.4 Prototype Develop and Test Evaluation

ZTE has already completed the mMTC BS and UE prototype development based on MUSA and started on-site tests. Some tests have already been completed in a test site in Shanghai, with the specific test deployment illustrated in Figure 3.24.

In Figure 3.24, the left figure is the actual test site, which includes 12 UEs. Each group containing four UEs, is randomly placed and marked with solid circle. A BS is deployed on the left and marked with dashed circle. The right figure describes the work principle of on-site tests briefly. The server distributes data packets with different size through gateway to each UE. The processing results of the BS and the throughput test results are displayed on the BS monitor software through the gateway, so that the testers can deal with these results. ZTE mainly tested the uplink high overload capability of MUSA. The test results are illustrated in Figure 3.25.

The test results of Figure 3.25 is obtained in a unit time on physical data channel with 4 PRBs; 12 UEs transmit information on 4 PRBs simultaneously. The multiplex scheme of MUSA on fixed frequency band is illustrated in Figure 3.26. In Figure 3.25, the first column gives the amount of received data at the BS, while the following 12 columns give the amount of transmitted data of each UE. Due to possible data collisions, actual amount of transmitted data of some BSs is low. From the test results it is obvious that the access user number of MUSA is 300% of LTE, and the uplink throughput is close to 300% of LTE.

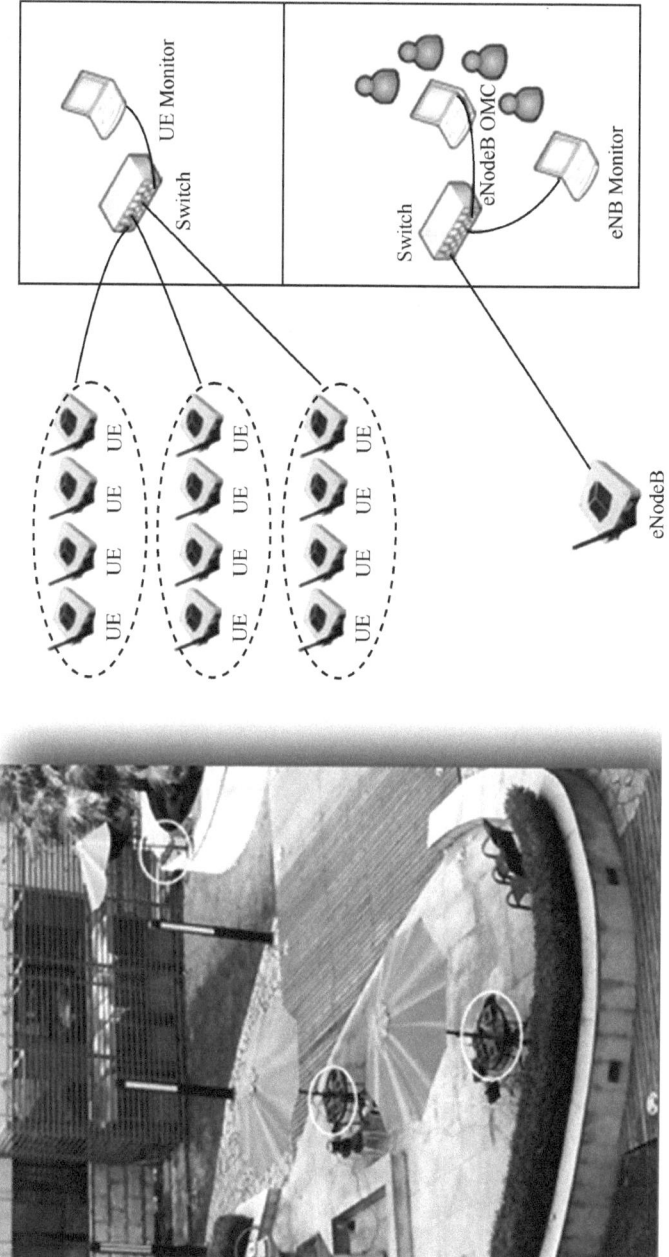

Figure 3.24: Test deployment of MUSA.

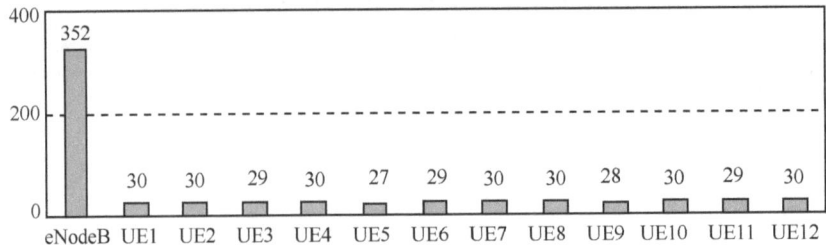

Figure 3.25: Test case-high overload (300% SISO).

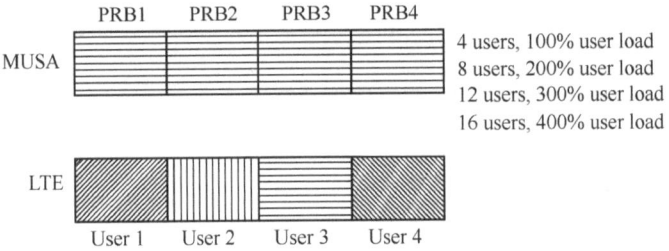

Figure 3.26: Complex comparison between MUSA and LTE in multiple user PRBs test scenario.

3.4 NOMA

Xin Su

3.4.1 Technical Principle

NOMA is an NMA technology based on power allocation, which realizes the transmission on same time, frequency or space domain resources by simple linear superposition of signals from multiple users and separates different user signals by applying advanced receivers (such as SIC receiver). The uplink and downlink capacity of NOMA systems can thereby approach the capacity bound.

Take two users as an example; Figure 3.27 shows the downlink signal processing flow of transmitter and receiver in NOMA systems [2]. User 1 is a cell-center user, while user 2 is a cell-edge user. At the transmitter, user 1 and user 2 occupy same time/frequency/space domain resources, and signals from the two users superpose in power domain. The BS allocates low power to user 1 with good channel conditions, while allocates high power to user 2 with poor channel conditions. The superposed signal transmitted by the BS can be expressed as:

$$x = \sqrt{\beta_1} x_{UE1} + \sqrt{\beta_2} x_{UE2} \tag{3.13}$$

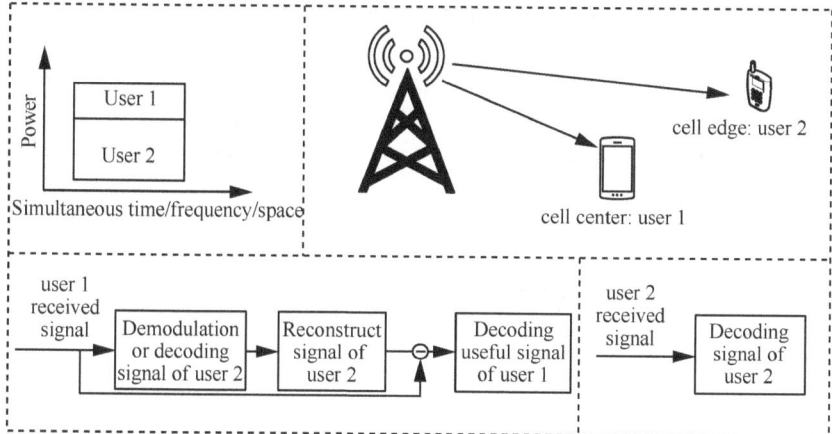

Figure 3.27: Principle of NOMA.

where x_{UE1} and x_{UE2} denote the signal of user 1 and user 2, respectively, β_k is the power allocation factor of user k (k=1,2), satisfying $\beta_1 + \beta_2 = 1$. To separate signals from different users at the receiver, the power allocation factors of the two users should not equal to each other, that is $\beta_1 \neq \beta_2$. Besides, the sum of power that the BS allocates to the users is limited by the maximum transmission power of the BS.

In the signal received by user 1, the power that the BS allocated to user 2 is larger than the power of user 1. To guarantee the correct demodulation of the signal of user 1, the signal of user 2 needs to be demodulation/decoded and reconstructed first, then the reconstructed signal of user 2 is subtracted from the received super-posed signal. By eliminating the signal interference of user 2, the signal of user 1 can be decoded under good SINR conditions, thus the correct decoding rate of user 1 is improved. When the signal of user 2 is completely subtracted, the relation between the received SINR of user 1 in NOMA and OMA system can be expressed as:

$$SINR_{NOMA, UE1} = \beta_1 \cdot SINR_{OMA, UE1} \tag{3.14}$$

where $SINR_{NOMA, UE1}$ and $SINR_{OMA, UE1}$ denote the SINR of user 1 in NOMA and OMA system, respectively.

Although there is signal interference of user 1 in the received signal of user 2, the interference power of user 1 will be less than the signal power of user 2 since high power is allocated to user 2 at the transmitter. By appropriate Adaptive Modulation and Coding (AMC) and treating the interference of user 1 as noise, the signal of user 2 can still be decoded correctly. The relation between the received SINR of user 2 in NOMA and OMA system can be expressed as:

$$SINR_{NOMA, UE2} = \beta_2 / \left(\beta_1 + \frac{1}{SINR_{OMA, UE2}} \right) \tag{3.15}$$

where $SINR_{NOMA,UE2}$ and $SINR_{OMA,UE2}$ denote the SINR of user 2 in NOMA and OMA system, respectively [25].

By signal superposition in power domain, NOMA can obtain a high multiplex gain, which improves the spectrum efficiency and the system capacity. NOMA can be well combined with multi-antenna technology. It is also well compatible with OFDMA and SC-OFDM access.

3.4.2 Transmission Scheme and Key Technology

1. Transmit Scheme and Key Technology

The user number to be supported by actual NOMA systems is far more than 2. Without loss of generality, we suppose there are M users in a cell, and the bandwidth in the downlink is divided into multiple sub-bands. In each sub-band, the BS superposes signal from multiple users and transmits through downlink channel to users. Denote the user number that one sub-band can carry by N. The user numbers that different sub-bands can carry may not the same. Denote the maximum user number that one sub-band can carry simultaneously by $N_{\max}$ [26].

Suppose the N users simultaneous carried in one sub-band are listed in descending order of SINR (user 1 has the highest SINR, user N has the lowest SINR), then the received signal of user n can be expressed as:

$$y_n = h_n \sum_{k=1}^{N} s_k \sqrt{P_k} + I_n + n_n \tag{3.16}$$

where $n \in [1,N]$, h_n denotes the channel response from the BS to user n, s_k is the transmitted signal of user k, P_k denotes the power of user k, with $P_k = \beta_k P_{BS}/N_{SB}$ ($\beta_k \in (0,1)$ is the power allocation factor of user k, P_{BS} is the total transmit power of the BS, N_{SB} is the sub-band number in the system). The inter-cell interference and AWGN received by user n are denoted by I_n and n_n, respectively.

The power allocation factor of a cell-edge user is bigger than the power allocation factor of a cell-center user. SIC needs to handle users in ascending order of SINR to achieve the best performance. The $SINR_n^{Post}$ after SIC can be obtained by Eq. (3.18).

$$y_n = h_n s_n \sqrt{P_n} + h_n \sum_{k=n+1}^{N} s_k \sqrt{P_k} + h_n \sum_{k=1}^{n-1} s_k \sqrt{P_k} + I_n + n_n \tag{3.17}$$

$$SINR_n^{Post} = \frac{\beta_n}{\sum_{k=1}^{n-1} \beta_k + \frac{1}{SINR_n}} \tag{3.18}$$

where $SINR_n = |h_n|^2 P_{BS}/(N_{SB}P_{1+N})$ is the SINR when maximum user transmission power is used, P_{1+N} is the total power of inter-cell interference and noise [27].

(1) Power Allocation

The NOMA system deals with power allocation mainly through setting suitable power allocation factors. From Eq. (3.18), it can be seen that, power allocation factors decide the $SINR_n^{Post}$ after SIC detection and the applied MCS for each user data. By adjusting the power allocation factors, the BS can control the throughput of each user freely. The throughput and the fairness of users in the cell are also tightly related to the power allocation factors.

A power allocation scheme is introduced here based on the average user throughput maximization criterion. The average user throughput maximization can be expressed as:

$$\{\beta_1^*, \beta_2^*, \cdots, \beta_N^*\} = \arg\max\left\{\sqrt[N]{\prod_{n=1}^{N} SE_n}\right\} \tag{3.19}$$

$$SE_n = SE_n^{MCS^*}\left(1 - BLER_n^{MCS^*}\right)\Big|_{SINR_n^{Post}} \tag{3.20}$$

$$MCS^* = \arg\max_{\{MCS\}}\left(SE_n^{MCS}\left(1 - BLER_n^{MCS}\right)\Big|_{SINR_n^{Post}}\right) \tag{3.21}$$

where $\beta_n \in (0,1)$, and $\sum_{i=1}^{N}\beta_i = 1$; SE_n is the achievable spectrum efficiency of user n; SE_n^{MCS} and $BLER_n^{MCS}$ denote the spectrum efficiency and BLER with chosen MCS, respectively, and $SINR$ equals to $SINR_n^{Post}$ in this case. More specifically, we can obtain $SINR_n^{Post}$ from (3.18) and then detect each MCS in the MCS set for user n. From the BLER and SINR curves in link-level simulations, it can be found that there is a one-to-one correspondence between $BLER_n^{MCS}$ and $SINR_n^{Post}$. Next, by computing $SE_n^{MCS}(1 - BLER_n^{MCS})$, candidate achievable spectrum efficiency for all MCS is obtained, and the largest one in the candidate achievable spectrum efficiency set is denoted by MCS^*. By adopting this power, allocation between overall user throughput and user fairness can be realized; however, the computing complexity increases exponentially with the growth of the multiplexed user number.

(2) User Schedule

In order to choose the best user set from all possible user sets, user schedule algorithm and power allocation algorithm need to be jointly used. Take the schedule algorithm based on proportional fairness in multi-user transmissions as an example. To achieve a higher overall user throughput, schedule algorithm can maximize the proportional fairness of multi-user transmissions, which is similar with the single user proportional fairness algorithm in terms of user fairness.

The BS schedules for each sub-band, and all possible user sets in each sub-band can be expressed as:

$$\binom{M}{1} + \binom{M}{2} + \cdots + \binom{M}{N_{\max}} \tag{3.22}$$

For each user set, power allocation algorithm is used to decide the power allocation rate of the users in the set. In order to select the best user set in the kth sub-band, appropriate metric needs to be applied. User set metric is defined as:

$$\Lambda_{\psi_j} \triangleq \prod_{n \in \psi_j} \left(1 + \frac{r_{k,n}(t)}{(t_c - 1)R_n(t)} \right) \tag{3.23}$$

where ψ_j denotes the jth candidate user set, $r_{k,n}(t)$ denotes the instant throughput of user n in the kth sub-band at time instant t, $R_n(t)$ denotes the average throughput of user n at time instant t, t_c is the size of the window used for computing the user throughput.

The best user set in one sub-band is denoted by $\psi^{Optimal}$, where $\psi^{Optimal} = \arg \max \psi_j \left(\Lambda_{\psi_j} \right)$. In reference [28], this schedule metric is proved to be the best metric for multi-user transmission. After scheduling, the average user throughput of all sub-bands can be updated by the following formula.

$$R_n(t+1) = \left(1 - \frac{1}{t_c} \right) R_n(t) + \frac{1}{t_c} \sum_{k=1}^{N_{SB}} r_{k,n}(t)/N_{SB} \tag{3.24}$$

The above formula indicates the average throughput updates slide window based on a window size t_c. If user n is not scheduled in the kth sub-band, then $r_{k,n} = 0$ [27].

2. NOMA Receiver Design

NOMA applies SIC technology, which includes SL-SIC and CW-SIC. In the SL-SIC receiver, user 1 first demodulates the symbol-level signal of user 2, then subtracts the signal of user 2 in order to obtain its own signal [25]. The CW-SIC receiver adds decoding and re-encoding modules based on the SL-SIC receiver. In CW-SIC, the information of a cell-edge user after demodulation and decoding implements re-encoding and modulation to reconstruct the signal of the cell-edge user, then the reconstructed signal of the cell-edge user is subtracted from the received signal.

The comparison of link-level performance between CW-SIC receiver and SL-SIC receiver in NOMA systems is given in reference [27]. The simulation results show that CW-SIC has a similar performance with ideal SIC receiver, while there is a wide performance gap between SL-SIC and ideal SIC. Compared with SL-SIC, CW-SIC improves the BLER performance and suppresses error propagation to a certain degree. However, decoding and encoding increases the computing complexity and detect latency of the receiver [27].

3.4.3 Simulation Evaluation and Performance Analysis

1 Link-Level Simulation and Performance Analysis

Reference [2] compares the BLER performance of NOMA with different receivers when the downlink power allocation proportion is {0.8, 0.2} and {0.7, 0.3}. Main simulation parameters are listed in Table 3.10, and the simulation results are illustrated in Figure 3.28.

Table 3.10: Downlink link-level simulation parameters.

Parameter	Value
Waveform	OFDM
System bandwidth	9 MHz
Antenna configuration	BS: 1 Tx UE: 2 Rx
Subcarrier spacing	15 kHz
Subcarrier number	600 (50 RB)
FFT size	2048
Subframe length	1.0 ms (14 OFDM symbols)
Symbol duration	66.67us + CP: 4.69us
Modulation scheme of cell-center user	16QAM
Modulation scheme of cell-edge user	QPSK
Channel coding/decoding	Turbo coding/Max-Log-MAP decoding
Channel model	7 path EPA
Maximum Doppler shift	5.5 Hz
FFT timing detection	Ideal
Channel estimation	Ideal channel estimation

From the simulation results, it can be seen that CW-SIC receiver has a similar performance with ideal SIC receiver. There is a wide performance gap between SL-SIC receiver and ideal SIC receiver, especially when the power of the cell-edge user is high. This indicates that the effect of error propagation cannot be ignored in SL-SIC receiver.

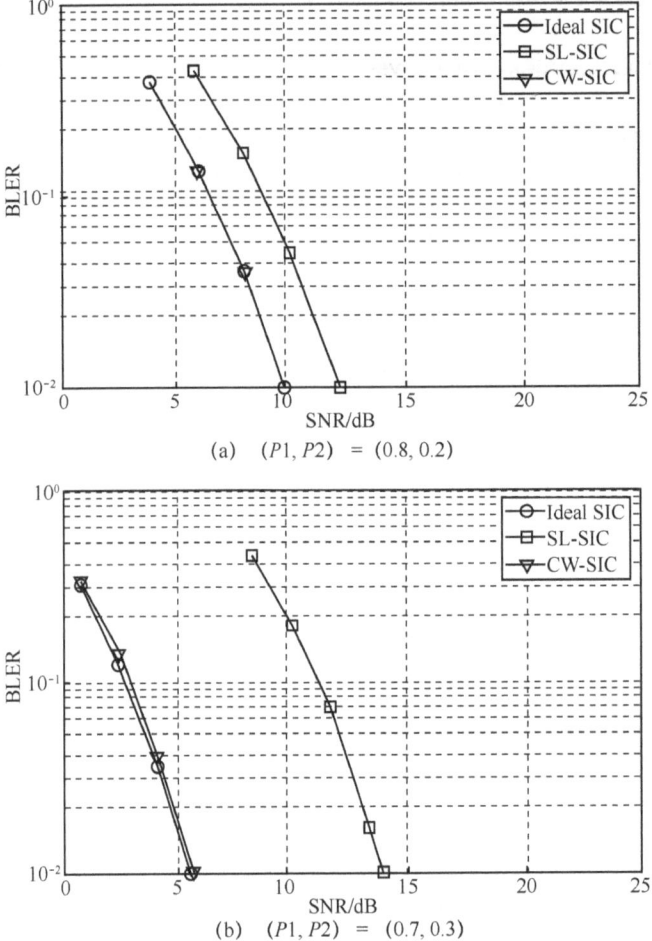

(a) $(P1, P2) = (0.8, 0.2)$

(b) $(P1, P2) = (0.7, 0.3)$

Figure 3.28: Performance of different receivers in two-user NOMA system.

2 System-Level Simulation and Performance Analysis

(1) Uplink System-Level Simulation

The main parameters of NOMA uplink system-level simulation are listed in Table 3.11 [26]. The locations of the UEs are randomly generated with a uniform distribution within each cell. The same MCS sets are used for both SC-FDMA and NOMA in the simulations. We consider inter-cell interference coordination with and without Fractional Frequency Reuse (FFR) in NOMA systems in the simulation. In FFR evaluations, 16 resource blocks are defined as edge bands for each cell, which are non-overlapped among the 3 adjacent cells. Within each cell, 1/3 UEs out of the total UEs are categorized as cell-edge UEs based on power of their reference signal transmitting from the base station,

Table 3.11: Uplink system-level simulation parameters of NOMA.

Parameter		Value
Cell layout		Hexagonal 19-cell sites, 3 cells per site, wrap around
Inter-site distance		500 m
Carrier frequency		2.0 GHz
Overall transmission bandwidth		10 MHz
Resource block bandwidth		180 kHz
Number of resource blocks		48
Sub-band size		6 PRBs for without FFR; 8 PRBs for with FFR
Number of UEs per cell		10, 20, 30, 40, 50
eNB receive antenna	Number of antennas	2
	Antenna gain	14 dBi
UE transmit antenna	Number of antennas	1
	Antenna gain	0 dBi
Maximum transmission power		23 dBm
Distance dependent path loss		$128.1 + 37.6\log_{10}(r)$, r. kilometers (dB)
Shadowing standard deviation		8 dB
Channel model		6-ray Typical Urban
Channel estimation		Ideal channel estimation
Receiver noise density		−174 dBm/Hz
Noise figure of cell site		5 dB
UE speed (Doppler frequency)		3 km/h (5.55 Hz)
Scheduling interval		1 msec
Average interval of throughput		200 ms
Traffic model		Full buffer

while others are cell-center UEs. Both the average UE throughput and cell-edge UE throughput are evaluated, where the cell-edge UE throughput is defined as the 5% value of the cumulative distribution function of the UE throughput.

Figure 3.29 presents the overall cell throughput of SC-FDMA and NOMA without FFR. The sub-band number is set to 8 and the maximum multiplexing order $N_{\max}$ for NOMA is set to 2. It can be seen that cell throughput of SC-FDMA is almost saturation when the number of UEs is larger than 10. However, when the number of UEs per cell is larger than 40, cell throughput of NOMA still increases as number of UEs

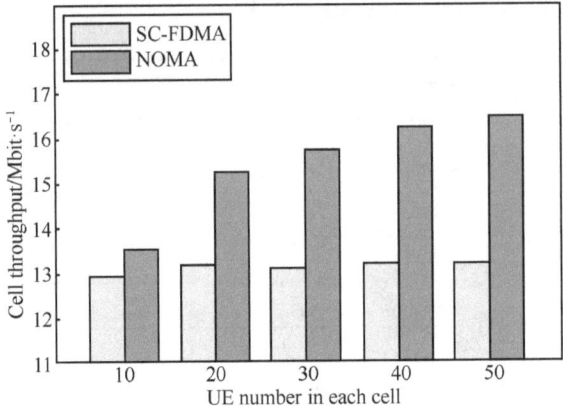

Figure 3.29: Comparison of cell throughput of SC-FDMA and NOMA without FFR.

per cell becomes larger. NOMA achieves about 28% cell throughput gain compared with SC-FDMA. The gain mainly comes from the non-orthogonal multiplexing, which substantially improves the resource utilization efficiency by transmitting multi-user data on one RB.

Figure 3.30 compares the UE throughput of SC-FDMA and NOMA with and without FFR when N_{max} is set to 3. It can be seen that: 1) the cell-edge user throughput of NOMA with FFR is better than SC-FDMA; 2) NOMA with FFR improves not only cell-edge throughput gain but also overall cell throughput gain [26].

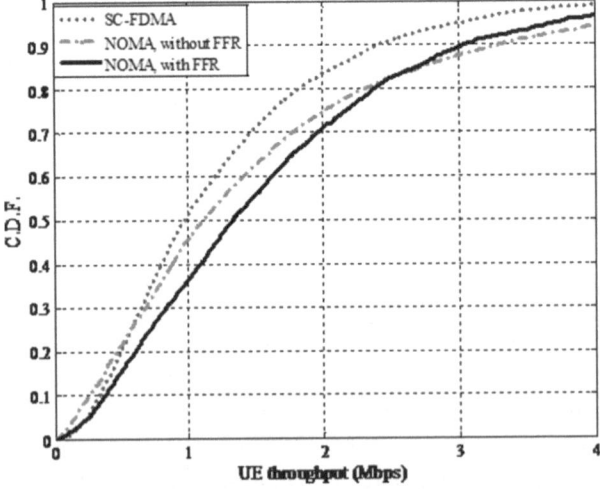

Figure 3.30: UE throughput cumulative distribution function.

(2) Downlink System-Level Simulation

2×2 antenna configuration and closed loop transmission mode 4 are applied in NOMA downlink system-level simulations. Detailed simulation parameters are listed in Table 3.12. The Coupling Loss (CL) between user k and the serving cell is denoted by CL_k, which is the power difference between the transmitter and the receiver and includes large scale fading and the antenna gain. A user can be identified as a cell-center or a cell-edge user based on certain CL threshold CL_{thre}: if $CL_k \leq CL_{thre}$, user k will be identified as a cell-center user; if $CL_k > CL_{thre}$, user k will be identified as a cell-edge user. Each candidate NOMA user set includes one cell-center user and one cell-edge user.

Table 3.12: Downlink system-level simulation parameters.

Parameter	Value
Layout	Hexagonal 19-cell sites, 3cells per site
Inter-site distance	500 m
Overall transmission bandwidth	10 MHz
Carrier frequency	2.0 GHz
Maximum BS transmit power	46 dBm
Channel model	ITU Uma
BS antenna pattern	3D
BS antenna height	25 m
BS antenna gain	17 dBi
UE antenna height	1.5 m
UE antenna gain	0 dBi
Antenna configuration	2 Tx/2 Rx
Traffic model	Full buffer
Scheduled UE number in NOMA	2
UE noise coefficient	9 dB
UE speed	3 km/h
Cell selection mechanism	Reference Signal Received Power，RSRP
Transmission mode	SU-MIMO closed loop transmission mode 4; multi-user superposition in power domain
CSI feedback assumption	Feedback period: 5 ms; feedback latency: 6 ms

In NOMA systems, the choice of user power allocation factor β is very important. In the simulation, candidate power allocation factor set including $\beta = \{0.1, 0.2, 0.3, 0.4\}$ is pre-defined, and the best β is chosen for the user pair according to the PF criterion. Besides, based on the multi-user PF criterion, the system platform supports adaptive switching between OMA and NOMA.

Channel Quality Indication (CQI) and Precoding Matrix Indicator (PMI) feedback enhancement are not considered in the simulation. Existing OMA CQI is updated for the estimation of NOMA CQI.

Consider two simulation cases: ten or twenty users in each cell on average. When there are ten users in each cell, NOMA can bring 13.4% cell average gain and 18.5% cell edge gain compared with OMA; when the user number increased to 20 in each cell, cell average gain and cell edge gain are 19.4% and 29.1%, respectively [2]. Whether the increase of cell-edge user number or increase of cell-center user number, the gain of average system increases, which indicates a better performance improvement is obtained by NOMA compared with OMA.

There are three obvious differences between NOMA and OFDMA: 1) NOMA has a better performance than OFDMA. In OFDMA, one sub-band can only be occupied by one user. While in NOMA, one sub-band can be occupied by multiple users, thereby the sub-band resource is fully utilized, which is the main reason that NOMA can achieve high performance; 2) The NOMA gain increases when the user number increases from 10 to 20, since the BS can find more suitable UEs for NOMA multiplexing when the UE number in the cell increases; 3) The NOMA gain is larger in wideband user pairing and scheduling than in sub-band. The performance gain is 13.4% for cell-edge throughput and 18.5% for cell-center throughput when the UE number is 10. When the UE number is 20, the NOMA gain is 19.4% for cell-edge throughput and 29.1% for cell-center throughput. The cell-center throughput gain is larger than the cell-edge throughput gain. The reason is that for wideband user pairing and scheduling, the wideband power allocation ratio can well match the wideband MCS while for sub-band user pairing and scheduling, the power allocation ratio is calculated on sub-band basis but the MCS is selected on wideband basis, which means there is a mismatch between the power allocation and MCS for transmission [27].

3.4.4 Prototype Develop and Test Evaluation

In order to verify the feasibility and the performance of NOMA in actual systems, such as the effect of actual SIC receiver (non-ideal characteristic of radio frequency hardware), the precision effect of analog-to-digital conversion and so on, NOMA verification platform as illustrated in Figure 3.31 is built for tests. The parameter configurations are: carrier frequency is 3.9 GHz, user bandwidth is 5.4 MHz (NOMA) or 2.7 MHz (OFDMA), frame structure of LTE Release 8, channel estimation based on Code Rate Spreading (CRS), 16 bit analog-to-digital conversion, CW-IC receiver applied

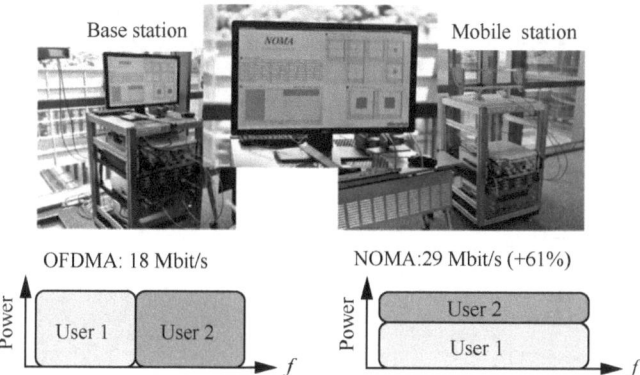

Figure 3.31: NOMA prototype verification.

by cell-center users, MMSE receiver applied by cell-edge users and cell-center users for MIMO detection, 2×2 matrix channel produced by channel simulator. The performance of NOMA is verified by using the platform with the above configurations, and 61% performance gain can be observed compared with OFDMA systems [2].

3.5 RSMA

Jie Zeng

3.5.1 Technical Principle

RSMA is a non-orthogonal multiple access technology by combining low-rate channel coding and scrambling code (or different interleavers), in which users are distinguished by scrambling code with good correlation characteristic [29]. As illustrated in Figure 3.32, RSMA can realize simultaneous transmission of multiple users with all time-frequency resources. In order to introduce the principle of RSMA more clearly, Figure 3.33 shows the transmitter-receiver structure of RSMA. At the transmitter, after Turbo coding,

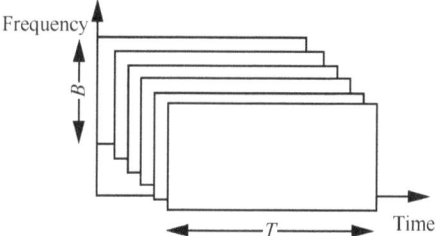

Figure 3.32: RSMA scheme.

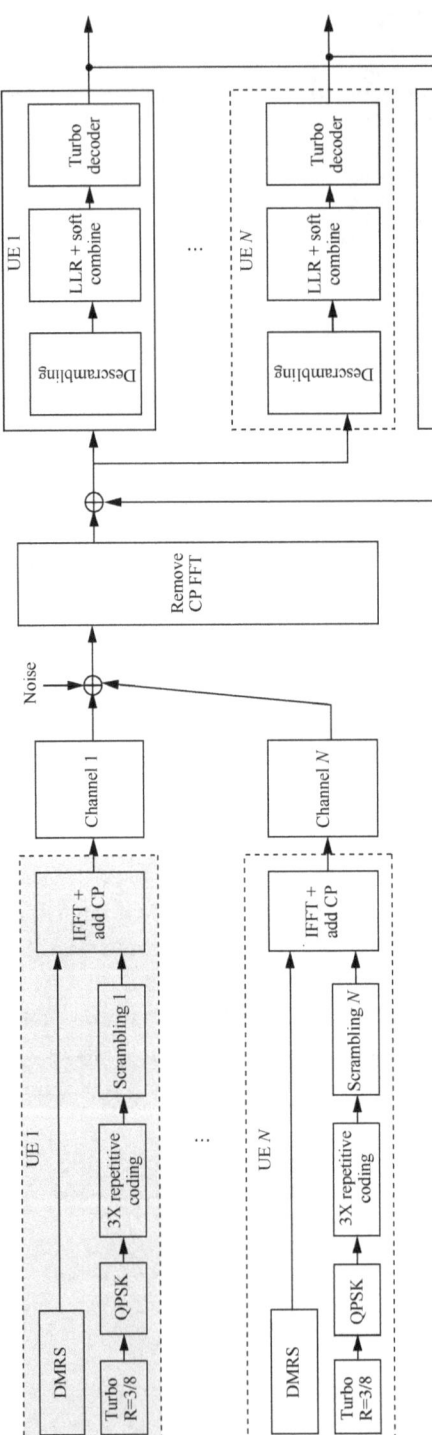

Figure 3.33: Transmitter-receiver structure of RSMA.

modulation and repetitive coding, signals of different users are scrambled by different scramble sequences and transmitted after OFDM modulation. At the receiver, after OFDM demodulation, SIC is used to separate the superposed multi-user signal.

3.5.2 Transmission Scheme and Key Technology

1. Transmit Scheme and Key Technology

The flexibility of RSMA is that it can combine different waveform and modulation schemes according to specific system design goals.

In the example illustrated in Figure 3.34(a), since link budget of UE is limited and battery saving needs to be considered, RSMA combine single carrier waveform. RSMA combined with grant-free transmission can reduce the signaling overhead effectively, meanwhile single carrier waveform can further decrease the Peak to Average Power Ratio (PAPR) to obtain a higher power amplifier efficiency. Pulse shaping module can further enhance the PAPR performance (such as constant envelope waveform) and reduce out-of-band emission.

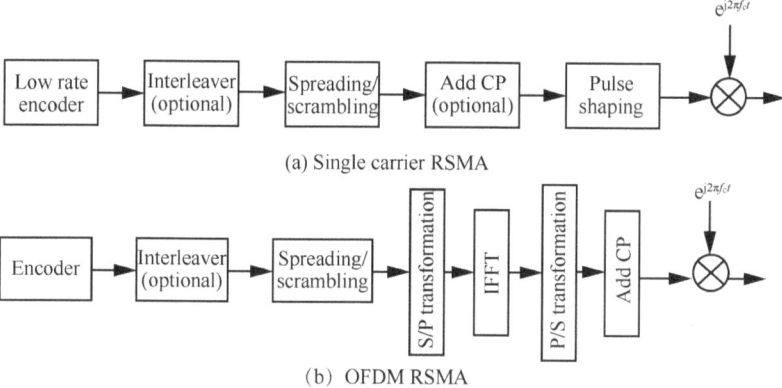

Figure 3.34: Combinations of RSMA and different waveforms.

In the example illustrated in Figure 3.34(b), to reduce the access latency becomes the first goal of system design. In this example, the UE is in connected mode and have synchronized with the BS, whose link budget is not limited (such as close to the BS). Such UE can apply RSMA with grant-free transmission and combine it with multi-carrier waveform based on OFDM to reduce the overall access latency.

2. Receiver Schemes and Key Technology

Figure 3.35 gives the RSMA receiver block diagram [30, 31]. In RSMA, following two multi-user detection algorithms with low complexity are used generally.

1) Match Filtering (MF): signal of each layer executes descramble and dispreading before it is sent to the Low Density Parity Check Code (LDPC) decoder. Multi-user detection is implemented by conjugate transpose of the spreading sequences/scrambling codes, which can be seen as the MF.

2) MF + SIC: after LDPC decoding of the data packet, decoded signal of this data packet is subtracted from the received signal by the receiver, then data packets that are not decoded successfully will be decoded again until all data packets are decoded.

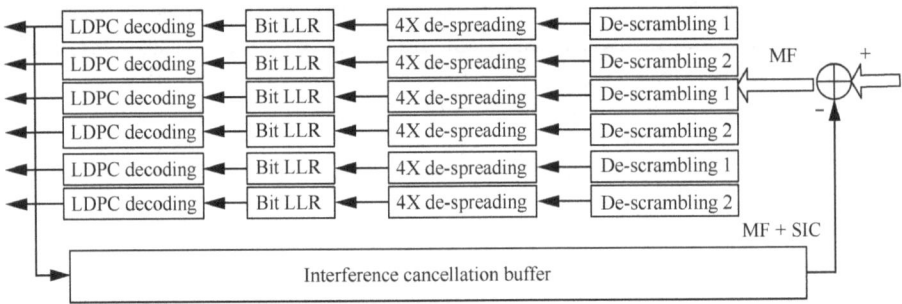

Figure 3.35: RSMA receiver block diagram.

3.5.3 Simulation Evaluation and Performance Analysis

1. Link-Level Simulation and Performance Analysis

For users with the same RSMA Grade of Service (GoS), ideally, the receiving power P_{rx} of each user at the BS should be the same (such as with the help of power allocation). To satisfy certain BLER requirement, received signal needs to satisfy a minimum SNR, which is:

$$\frac{P_{rx}}{N_t} = \frac{P_{rx}}{N_0 B + (K-1)(1-\beta)P_{rx} + fKP_{rx}} \geq \frac{ebno}{D} \cdot \frac{R}{B} \tag{3.25}$$

where $ebno$ is the needed bit-level SNR. It is important to note that $ebno$ is usually a function of spectrum efficiency per user $E_{spectra}$. From the Shannon formula, we have

$$ebno = \frac{2^{E_{spectra}} - 1}{E_{spectra}} \tag{3.26}$$

For RSMA, since signal of each user is spread to overall time and frequency resources, the spectrum efficiency of each user does not increase with change of the user number, and

$$E_{spectra} = \frac{R}{B} \left(bit/s \cdot Hz^{-1} \right) \tag{3.27}$$

Define link budget θ (dB) as the maximum permissible path loss, which can be expressed as:

$$\theta_{RSMA} = 10 \lg \left(P_{tx} G_{fixed} \right) - 10 \lg \left(\frac{ebno}{D} RN_0 \right) + 10 \lg \left(1 - \frac{ebno}{D} \cdot \frac{R}{B} (1 - \beta + f) \cdot K \right) \tag{3.28}$$

where G_{fixed} denotes cumulative effect of other fixed gains (received antenna gain, transmitted antenna gain, penetration loss, shadow and so on).

In the 3GPP RAN WG1 #86 meeting, Qualcomm proposed the performance comparison between RSMA and SCMA [31]. In the RAN WG1 #86b meeting, Qualcomm showed the simulation comparison between RSMA and OMA [32]. Link-level simulation parameter configuration is given in Table 3.13.

Table 3.13: Link-level simulation parameters.

Parameter	Value
Carrier frequency	2 GHz
Waveform	OFDM/SC-FDMA
Channel coding	LTE Turbo
System bandwidth	10 MHz
Total allocated bandwidth for transmission	4 PRB
Overhead	2 DM-RS symbols, no SRS
Target spectral efficiency	single UE spectral efficiency and multi-user complex are supported
BS antenna configuration	2, 4 Rx
UE antenna configuration	1 Tx
Transmission mode	TM1
Suggested SNR distribution of multiple UEs	Equal average SNR (short-term variation remains)
	Unequal average SNR (the SNR distribution is FFS)
Propagation channel & UE velocity	TDL-A 30 ns, 3 km/h
	TDL-C 300 ns, 3 km/h
Max number of HARQ transmission	1
Target BLER	0.1

Figure 3.36 provides the BLER versus SNR performance comparison between RSMA and OFDMA. SNR is defined as the received SNR per RE per Rx at eNodeB. Under equivalent average SNR conditions, Figures 3.36–3.38 provide performance comparison curves of RSMA and OFDMA with different spectrum efficiencies.

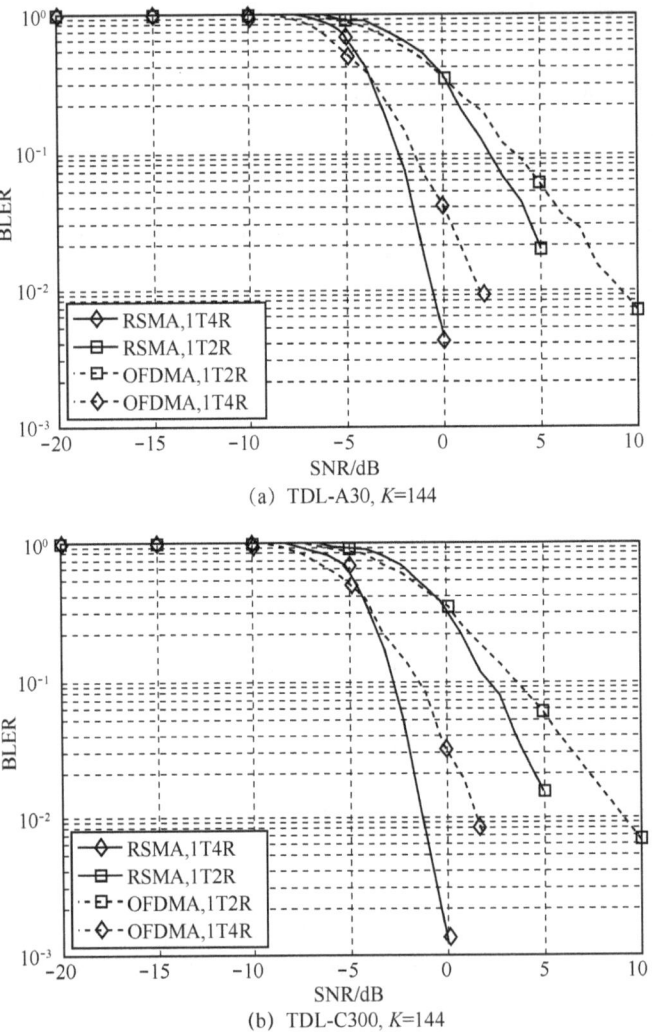

(a) TDL-A30, K=144

(b) TDL-C300, K=144

Figure 3.36: Spectrum efficiency of each user is 0.25 $bit \cdot s^{-1} \cdot Hz^{-1}$.

The two simulation diagrams in Figure 3.37 represent the BLER versus SNR performance comparison in different channels, respectively. From Figure 3.37, it can be seen that the SNR of RSMA is significantly lower than the SNR of OFDMA when the BLER is the same. Reference [32] also gives the performance simulation curves

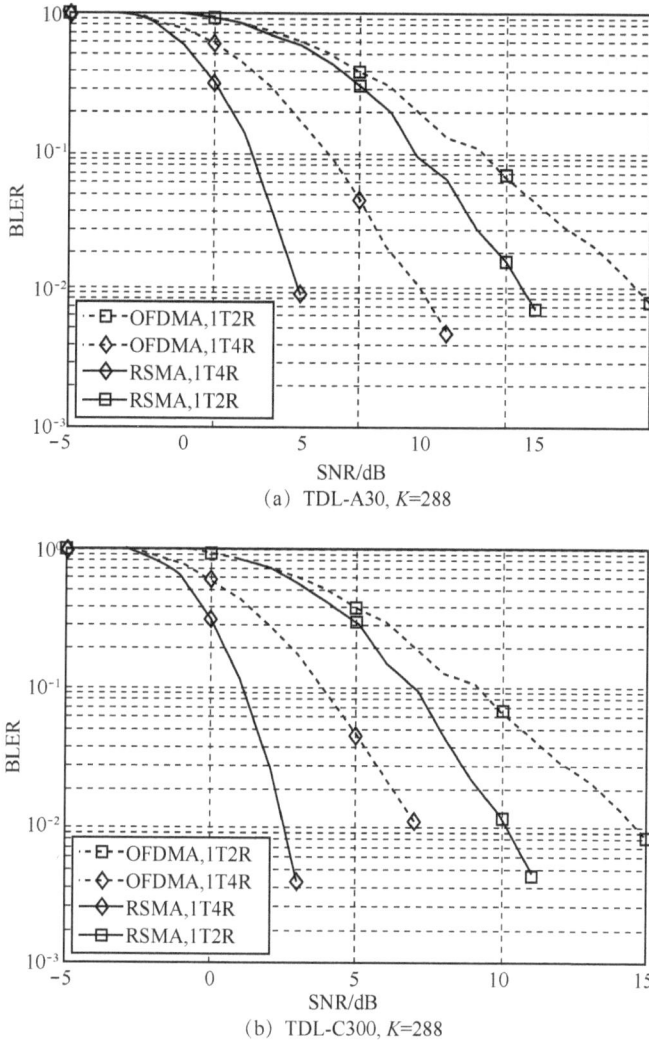

Figure 3.37: Spectrum efficiency of each user is 0.5 $bit \cdot s^{-1} \cdot Hz^{-1}$.

under nonequivalent average SNR conditions. RSMA has a more significant performance gain than OFDMA when the spectrum efficiency is high.

2. System-Level Simulation and Performance Analysis

References [33] and [34] give uplink system-level simulation results of RSMA in mMTC scenario.

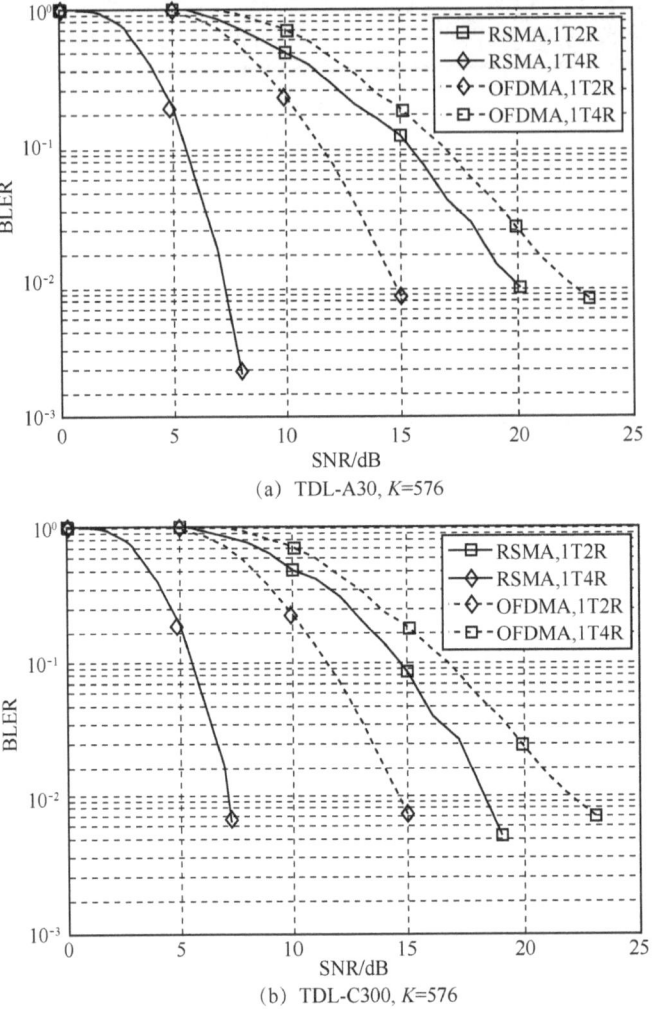

(a) TDL-A30, $K=576$

(b) TDL-C300, $K=576$

Figure 3.38: Spectrum efficiency of each user is 1.0 $bit \cdot s^{-1} \cdot Hz^{-1}$.

In this section, different link budget goals are used to evaluate system capacity of RSMA/FDMA/TDMA schemes. In order to explain the relations between link budget and different multiple access schemes, following case is considered: all users are symmetrical, which means they have same path loss and user GoS, and each user need to send 250 bit in every 10 ms, which corresponds to a 25 kbit/s normalized data rate. General simulation hypotheses are listed in Table 3.14.

Table 3.14: General hypotheses of system level-link budget.

Parameter	Value
Bandwidth	1.08 MHz
Receive antenna number	2 or 4
UE transmit power	0 dBm
UE transmit antenna gain	−1 dBi
Noise power spectrum density	−174 dBm/Hz
Inter-cell interference factor	0.5
Base station noise index	5 dB
Lognormal fading margin	6.9 dB
Penetration loss	20 dB
Base station receive loss (cable, connector)	2 dB
Base station receiving antenna gain	17 dBi
Shannon *ebno* fallback value	3 dB

Suppose there are two receiving antennas at the BS, Figure 3.39 compares corresponding link budget with different concurrent user number in a single cell. For the case of frequency division multiplexing and time division multiplexing, since each user

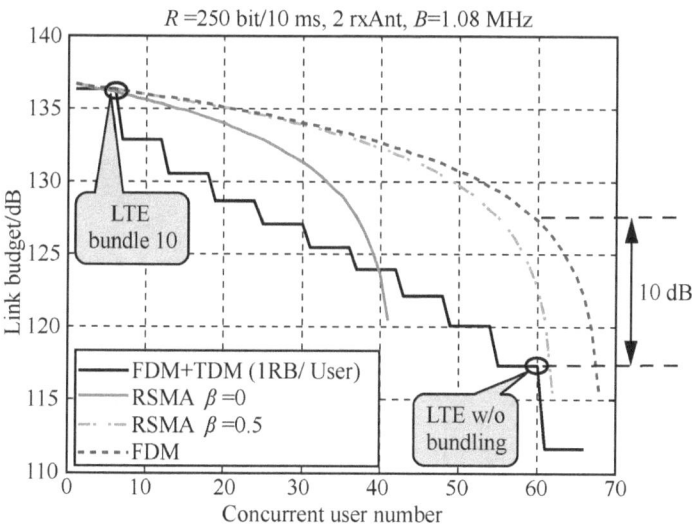

Figure 3.39: Link budget comparison (two receive antennas at the base station).

is allocated 180 KHz, the system can only support six users at the same time. When the user number is larger than six, time division multiplexing needs to be applied.

Since *ebno* is small, even interference cancellation is not used ($\beta = 0$), the link budget of RSMA (when capacity pole is achieved) is better than the multiple access system based on frequency division multiplexing and time division multiplexing. On the other hand, large intra-cell interference will finally cause low capacity pole. Specifically, when the last logarithmic term in Eq. (3.28) equals to zero, RSMA achieves the capacity pole.

$$K_{pole} = \frac{D \cdot B}{ebno \cdot R(1 - \beta + f)} \tag{3.29}$$

With the help of interference cancellation, the RSMA capacity pole can be greatly improved. As illustrated in Figure 3.39, when 50% of the intra-cell interference is cancelled ($\beta = 0.5$), RSMA has a comparable capacity pole with multiple access system based on frequency division multiplexing and time division multiplexing, meanwhile a good link budget is achieved. However, if arbitrarily small frequency resource can be allocated to each user, RSMA is still worse than frequency division multiplexing. It is worth noting that in practice, arbitrarily small frequency resource allocation method will cause lack of frequency diversity and let the system suffer from the effect of multi-path fading.

In order to explain the effect of receiving antenna number to link budget, the curves in Figure 3.39 apply the same configuration as Figure 3.40 except the receiving antenna number. With the increase of receiving antenna number, the performance gap between RSMA and multiple access system based on frequency

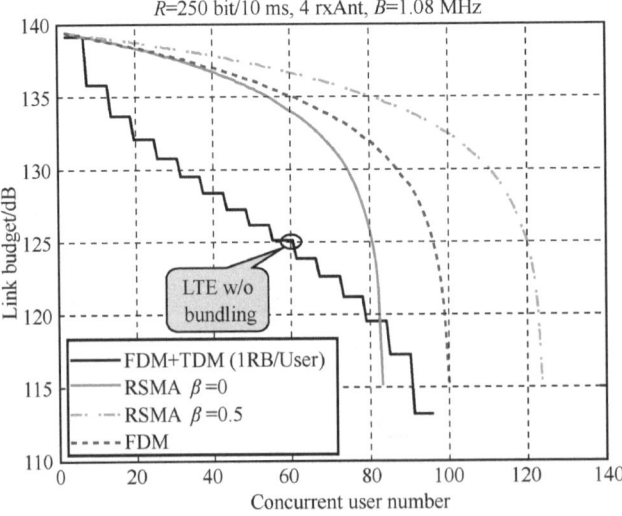

Figure 3.40: Link budget comparison (four receive antennas at the base station).

division multiplexing and time division multiplexing becomes wider. Meanwhile, whether interference cancellation is used or not, the capacity pole of RSMA multiplied. It is important to note that when interference cancellation is used ($\beta = 0.5$), the link budget performance of RSMA is better than frequency division multiple access.

3.6 IGMA

Jie Zeng

3.6.1 Technical Principle

IGMA is a NMA technology based on bit-level interleaving and grid mapping proposed by Samsung. The transmitter principle of IGMA is illustrated in Figure 3.41 [35].

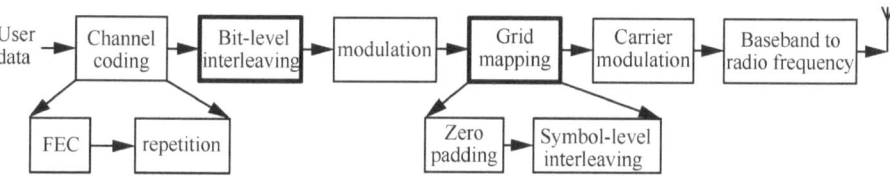

Figure 3.41: Transmitter principle.

After the processes including channel coding, bit interleaving, modulation and grid mapping, user data is modulated on carrier and transmitted through the baseband radio frequency. Channel coding can be implemented with simple repetitive coding or with low-rate Forward Error Correction (FEC) coding directly. Grid mapping includes zero padding and symbol-level interleaving. The concept of grid mapping pattern density ρ is introduced here. $\rho = N_{used}/N_{all}$ denotes the ratio of occupied RE number N_{used} to overall allocated RE number N_{all}, which can be configured flexibly.

Interleaving can be expressed mathematically as permutation of input data with permutation matrix $\mathbf{a}_I$. Define the coding bit sequence of user k as $\mathbf{b}_k = [b_{k,1}b_{k,2}b_{k,3} \cdots b_{k,M}]$, where M is the length of the coding bit sequence. After interleaving of coding bit sequence $\mathbf{b}_k$ and corresponding permutation matrix $\mathbf{a}_I$ with certain interleaver, interleaved bit sequence $\mathbf{C}_k = \mathbf{b}_k \cdot \mathbf{a}_I$ is obtained.

After interleaving, interleaved bit sequence is modulated to obtain the symbol sequence $\mathbf{S}_k = [S_{k,1}S_{k,2}S_{k,3} \cdots S_{k,N}]$ with length N. Then, the symbol sequence is sent to the gird mapping module.

Grid mapping module maps symbol to corresponding RE based on grid mapping pattern. The process can be expressed mathematically as permutation with matrix $\mathbf{a}_{GM}$. After zero padding and interleaving, the symbol sequence of user k is expressed as $\mathbf{S}'_k = \mathbf{S}_k \mathbf{a}_{GM} = [S'_{k,1} S'_{k,2} S'_{k,3} \cdots S'_{k,L}]$, where $L = N/\rho_k$, ρ_k decides zero padding number. For example, when $N = 4$ and $\rho_k = 0.5$, $L = 8$, the permutation matrix can be expressed as:

$$\mathbf{a}_{GM} = \begin{bmatrix} 0 & 0 & 0 & 0 & 0 & 0 & 0 & 1 \\ 0 & 0 & 1 & 0 & 0 & 0 & 0 & 0 \\ 0 & 0 & 0 & 0 & 0 & 1 & 0 & 0 \\ 1 & 0 & 0 & 0 & 0 & 0 & 0 & 0 \end{bmatrix} \qquad (3.30)$$

Then, the symbol sequence of user k after zero padding and symbol-level interleaving can be expressed as:

$$\mathbf{S}'_k = \mathbf{S}_k \mathbf{a}_{GM} = [S_{k,1} S_{k,2} S_{k,3} S_{k,4}] \cdot \begin{bmatrix} 0 & 0 & 0 & 0 & 0 & 0 & 0 & 1 \\ 0 & 0 & 1 & 0 & 0 & 0 & 0 & 0 \\ 0 & 0 & 0 & 0 & 0 & 1 & 0 & 0 \\ 1 & 0 & 0 & 0 & 0 & 0 & 0 & 0 \end{bmatrix} \qquad (3.31)$$

$$= [\, S_{k,4} \quad 0 \quad S_{k,2} \quad 0 \quad 0 \quad S_{k,3} \quad 0 \quad S_{k,1} \,]$$

After the above processes, original data is mapped to part of the allocated resource grid REs. In order to explain the concept of grid mapping more clearly, Figure 3.42 is used to describe this procedure.

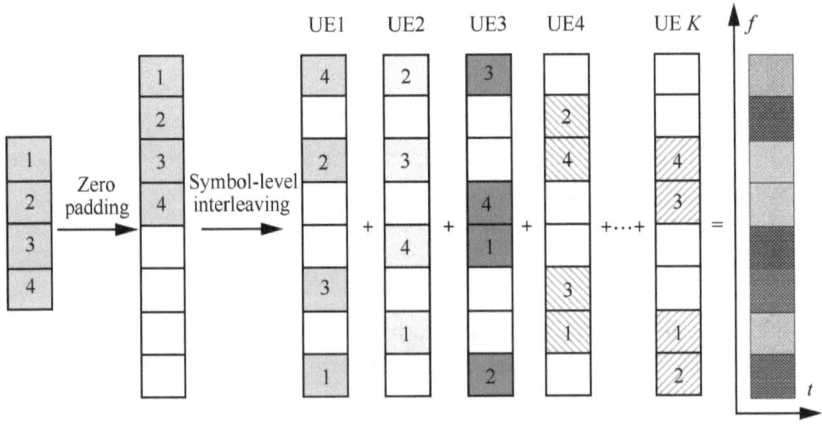

Figure 3.42: Grid mapping procedure with $N = 4$, $\rho_k = 0.5$, $L = 8$.

From the above, it can be seen that IGMA distinguishes users based on the following:
1) Different bit-level interleavers
2) Different grid mapping patterns
3) Combinations of bit-level interleavers and grid mapping patterns

Different connection density can be supported by IGMA with abundant bit-level interleavers and grid mapping patterns. Besides, by disorganizing the order of original symbol sequence with symbol-level interleaver, frequency selective fading and inter-cell interference can be resisted.

3.6.2 Transmission Scheme and Key Technology

1. Transmission Scheme and Key Technology

Design of the IGMA transmission scheme mainly includes bit-level design of interleavers and grid mapping patterns [36].

Interleavers can not only eliminate correlation among user chips, but also distinguish different users. In this section, several interleavers used in IGMA are introduced briefly.

(1) Random Interleaver

Random interleaver is one of the simplest interleave technologies, which has been widely used already. In random interleave scheme, interleaver of each user is randomly generated.

For any user k, suppose the interleaving depth of IGMA is N, the interleaver's generation is as follows.

1) Generate position index sequence $\{1, 2, \cdots, N-1, N\}$ with length N;
2) Set initial position index variable $i = 1$;
3) Generate position index increment j randomly, which satisfies $0 \leq j \leq N - i$;
4) Exchange position indexes i and $i + j$;
5) Compute $i = i + 1$;
6) If $i \leq N - 1$, then return to step 3); else stop loop.

When iteration is over, $i = N$. After iteration, the obtained new position index sequence is the interleaver of user k. The advantage of random interleaver is that the independence among interleavers of different users can be guaranteed well.

(2) Orthogonal Interleaver

Orthogonal interleaver implements with orthogonal sequences, thus can strictly satisfy orthogonality and decrease the interleaver design complexity. However, the interleaver number is limited by the length of the orthogonal sequence. When the user number is large, the application is greatly limited. Popular orthogonal sequences in orthogonal interleaver design include m sequence, Walsh sequence and so on.

(3) Pseudo-Random Interleaver

Since the number of orthogonal interleaver is limited by the length of the orthogonal sequence, non-orthogonal interleaver is needed when the user number is large. Pseudo-random interleaver is a low correlation interleaver proposed on this basis.

Pseudo-random interleaver generates pseudo-random sequences by using linear shift register defined by primitive polynomials.

Suppose the length of input information sequence is l, and $1/S$ repetitive code is used for encoding, thus the coded sequence length is lS. With the above conditions, pseudo-random interleaver design algorithm based on m sequence is obtained as follows:

1) Choose K m-order primitive polynomials, where the integer m must satisfy $2^m = lS$;
2) Generate K interleavers with length lS according to corresponding polynomials.

The procedures of generating interleavers according to corresponding polynomials is as follows:

1) Generate corresponding linear feedback shift register according to the coefficients of generator polynomial;
2) Set t ($1 \leq t \leq lS$) as discrete time, where $lS-1$ is the cycle length of m sequence, and t is initialized to $t=1$. Set $q_b(t)$ as the vector representation of shift register data at time t, and $q(t)$ is the decimal representation of $q_b(t)$.
3) In each m sequence (the cycle is $lS-1$), there exists longest continuous zero at a certain time x_0, which can be used to set the following interleave mapping rule:

$$\pi(t) = \begin{cases} q(t), & 1 \leq t \leq x_0 - 1 \\ lS, & t = x_0 \\ q(t-1), & x_0 + 1 \leq t \leq lS \end{cases} \tag{3.32}$$

Pseudo-random interleaver design is completed after the abovementioned steps. The advantage of pseudo-random interleaver is that it has low requirement to system resource, and the algorithm is simple.

(4) Nested Interleaver

The design ideas of nested interleaver are to complete the design of other interleavers in a nested way based on a basic primary pseudo-random interleaver. By re-interleaving the primary interleaver, nesting is implemented, and the other interleavers are obtained. The advantage of nested interleaver is the simple design and easy generation, meanwhile the implementation does not need too many storage resources.

Besides the above interleavers with mature design ideas, there are some other interleavers, such as shift interleaver, two-dimensional interleaver, three-dimensional matrix interleaver based on S-random interleaver, encrypted interleaver, particle swarm interleaver, symmetric interleaver, parallel interleaver, algorithm interleaver and so on. The interleaver designs in IGMA systems mainly evolves toward the direction of low complexity, fast generation speed, low correlation and large codeword distance.

The grid mapping pattern density in IGMA is related to the access user number, and directly affects the detection complexity at the receiver. By reasonable configuration of grid mapping pattern density, a balance between IGMA access user number and detection complexity at the receiver can be achieved.

2. Receiver Schemes and Key Technology

At the IGMA receiver, multi-user iterative detection algorithm based on ESE can be used to realize signal separation. Figure 3.43 illustrates the iterative multi-user detection. The outer information from decoder and the outer information from ESE iterate through interleaver and de-interleaver in order to separate multi-user signals.

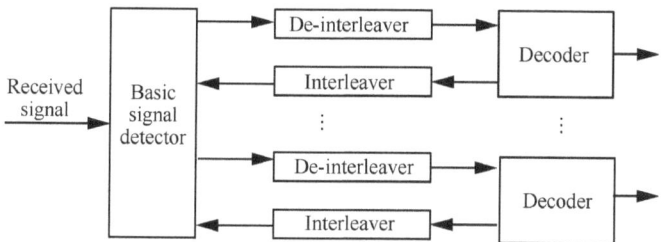

Figure 3.43: Iterative multi-user detection diagram.

Suppose the receiver and the transmitter are completely synchronized, and channel state information from user k ($k = 1, 2, \cdots, K$) to the receiver is h_k. The received signal at the receiver can be expressed as:

$$r^j = \sum_{k=1}^{K} h_k x_k^j + n^j \tag{3.33}$$

where n^j is the AWGN at time j, whose mean value is zero and variance is σ^2. The received signal can also be expressed as:

$$r^j = h_k x_k^j + \varsigma_k^j \tag{3.34}$$

where ς_k^j is the received information from the other $K - 1$ users and AWGN, which is:

$$\varsigma_k^j = \sum_{k' \neq k}^{K-1} h_{k'} x_{k'}^j + n^j \tag{3.35}$$

ς_k^j can be approximated to a Gaussian random variable.

Calculate the mean $E(r^j)$ and variance $Var(r^j)$ of the received signal. Suppose the mean and variance of the signal from the kth user are $E\left(x_k^j\right)$ and $Var\left(x_k^j\right)$, respectively, then

$$E(r^j) = \sum_{k=1}^{K} h_k E\left(x_k^j\right) \tag{3.36}$$

$$Var(r^j) = \sum_{k=1}^{K} |h_k|^2 Var\left(x_k^j\right) + \sigma^2 \tag{3.37}$$

Calculate the mean and variance of ς_k^j:

$$E\left(\varsigma_k^j\right) = E(r^j) - h_k E\left(x_k^j\right) \tag{3.38}$$

$$Var\left(\varsigma_k^j\right) = Var(r^j) - |h_k|^2 Var\left(x_k^j\right) \tag{3.39}$$

where the initial value of $E\left(x_k^j\right)$ is 0, the initial value of $Var\left(x_k^j\right)$ is 1.

After receiving the signal at the receiver, the signal is sent to ESE and we use log likelihood decoding. The log likelihood $e_{ESE}\left(x_k^j\right)$ of information sequence of user k after interleaving is:

$$
\begin{aligned}
e_{ESE}\left(x_k^j\right) &= \log\left(\frac{\Pr\left(x_k^j = +1 \middle| r^j\right)}{\Pr\left(x_k^j = -1 \middle| r^j\right)}\right) \\
&= \log\frac{\Pr\left(r^j \middle| x_k^j = +1\right)}{\Pr\left(r^j \middle| x_k^j = -1\right)} \\
&= \log\left(\frac{\exp\left(-\dfrac{\left(r^j - E\left(\varsigma_k^j\right) - h_k\right)^2}{2Var\left(\varsigma_k^j\right)}\right)}{\exp\left(-\dfrac{\left(r^j - E\left(\varsigma_k^j\right) + h_k\right)^2}{2Var\left(\varsigma_k^j\right)}\right)}\right) \\
&= \frac{2h_k\left(r^j - E\left(\varsigma_k^j\right)\right)}{Var\left(\varsigma_k^j\right)} \\
&= 2h_k \frac{r^j - E(r^j) + h_k E\left(x_k^j\right)}{Var(r^j) - |h_k|^2 Var\left(x_k^j\right)}
\end{aligned}
\tag{3.40}
$$

The outer information from ESE $e_{ESE}\left(x_k^j\right)$ is de-interleaved and considered as a priori log likelihood ratio to input to the decoder. The outer information from the decoder is feedback to ESE as a priori information, then new outer information estimation is calculated by ESE. The complexity of ESE is low. When the user number is K and Gray-Mapped order is $2Q$, the complexity is proportional to K and Q.

CBC MAP has a good performance. However, the complexity is high, which is proportional to $(2Q)^2$.

3.6.3 Simulation Evaluation and Performance Analysis

1. Link-Level Simulation and Performance Analysis

In the 3GPP RAN WG1 #86 meeting, Samsung gave link-level simulation results of IGMA [37]. In the simulation, the received power per RE and the spectrum efficiency per UE are need to be same for both IGMA and OFDMA. To guarantee fairness, the distributed resource allocation manner is adopted in OFDMA so that the frequency diversity gain can be achieved. Other specific simulation parameters are listed in Table 3.15.

Table 3.15: Link-level simulation parameters.

Parameter	Value
Carrier Frequency	2 GHz
Waveform	OFDM
Channel coding	LTE Turbo
Numerology	Same as Release 13
System Bandwidth	10 MHz
Total allocated bandwidth for transmission	6RBs (1.08 Mhz)
Overhead	2 DM-RS symbols, no SRS
Target spectral efficiency & supported UE number	Per UE SE = TB size/(144*RB number*grid mapping pattern density) ESE detector: 1) TB size (without CRC):{ 62, 84, 120, 148, 165, 192, 228, 264, 300} (bit); 2) grid mapping pattern density = 0.5 3) Supported UE numbers: {6, 8, 10, 12, 14, 16 } (equivalent overload rate: {150%, 200%, 250%, 300%, 350%, 400%}) CBC MAP detector: 1) TB size (without CRC):{84, 120, 192, 264, 408, 624 } (bit); 2) grid mapping pattern density = 0.5 3) Supported UE numbers: {6, 12} (equivalent overload rate: {150%, 300%})
BS antenna configuration	2 Rx

Table 3.15 (continued)

Parameter	Value
UE antenna configuration	1 Tx
Channel estimation	Ideal channel estimation
SNR distribution of Multiple UEs	Equal average SNR
Propagation channel & UE velocity	TDL-A & TDL-C; 3 km/h
Detection algorithm	ESE detector, CBC MAP detector
Target BLER	0.1

Figures 3.44 and 3.45 give the BLER comparison and the throughput comparison between IGMA and OFDMA, respectively. From Figure 3.44, it can be seen that when BLER is 10^{-2}, IGMA has a 1 dB gain compared with OFDMA whose MCS equals QPSK 1/2 and a gain near 2 dB compared with OFDMA whose MCS equals 16QAM 1/4. From Figure 3.45, it can be seen that IGMA can outperform OFDMA in terms of system throughput. When SNR equals to 12 dB, the throughput gain is about 40%. Combining Figures 3.44 and 3.45, it can be seen that IGMA has better throughput gain and BLER performance than OFDMA.

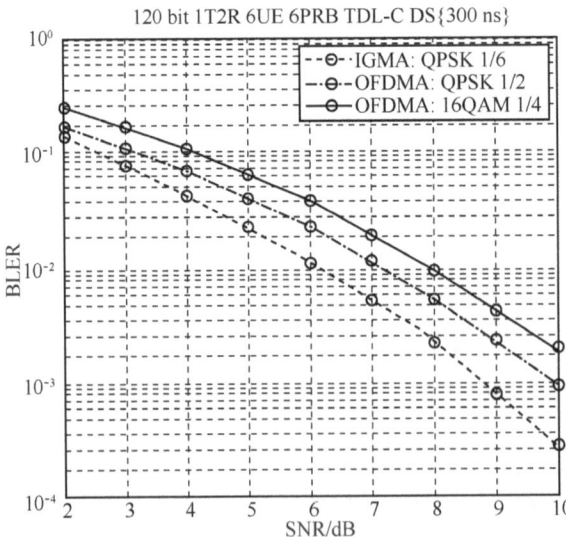

Figure 3.44: BLER performance comparison.

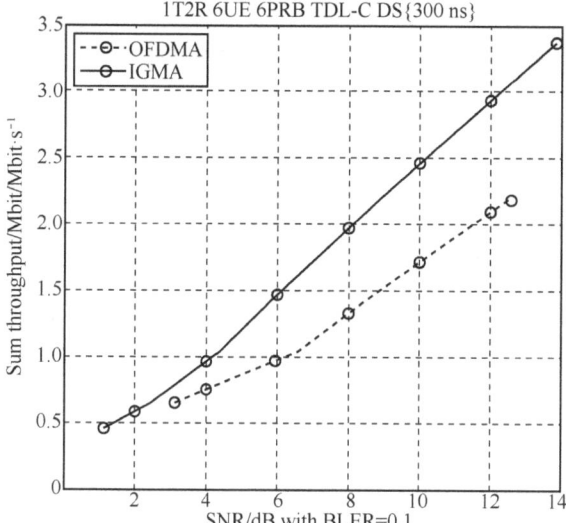

Figure 3.45: Throughput performance comparison.

2. System-Level Simulation and Performance Analysis

In the 3GPP RAN WG1 #86 meeting, Samsung concluded the IGMA system-level simulation results, including Physical Layer (PHY) abstraction and the calculation of system-level simulation metric [38]. For evaluation method, three aspects including PHY abstraction, metric calculation and further considerations on evaluation assumptions are proposed.

For mMTC scenario, the connection density should be used as the metric of non-orthogonal multiple access system-level evaluation [39]. Generally, the connection density can be calculated by pre-defined system packet loss rate. Define packet arrival rate λ as the per UE average inter-packet arrival time. Assuming the packet arrival rate for mMTC scenario is λ_{mMTC}, and the number of UE per cell is N_{UE}. The maximal packet arrival rate for given system packet loss rate threshold $p_{SLS}\%$ is λ_{SLS}, then the connection density is computed by

$$\rho_{mMTC} = \frac{N_{UE}\lambda_{SLS}}{\lambda_{mMTC}} \frac{1}{S_{\sec tor}} \tag{3.41}$$

where $S_{\sec tor}$ is the area of each cell.

The system-level simulation parameters of IGMA are listed in Table 3.16 [40].

The user number in each cell is 20. For IGMA, grant-free transmission is considered, the simulation bandwidth of each user is 6 PRBs, which is the same with the minimal bandwidth for mMTC. OFDMA with grant-free transmission is also considered, and each user chooses 1 PRB randomly from 6 PRBs. If two or more users choose the same PRB simultaneously for data transmission, collision happen. Since

Table 3.16: System-level simulation parameters.

Parameter	Value
Layout	Single layer–macro layer hexagonal grid
Inter-BS distance	1,732 m
Carrier frequency	700 MHz
Simulation bandwidth	6 PRBs
Channel model	3D UMa
Transmitting power	UE: Max 23 dBm
BS antenna configuration	Rx: 2 ports
Antenna element configuration	$(M, N, P, M_g, N_g) = (10, 1, 2, 1, 1)$, $d_v = d_h = 0.5\lambda$
Port mapping method	$(M_{tx}, N_{tx}, P, M_g, N_g) = (1, 1, 2, 1, 1)$, the mapping method following TR36.873
BS antenna height	25 m
BS antenna tilt	$6°$
BS antenna element gain + connector loss	8 dBi, including 3 dB cable loss
BS receiver noise coefficient	5 dB
UE antenna	1 Tx
UE antenna height	Follow the modeling of TR36.873
UE antenna gain	−4 dBi
Traffic model	Modified FTP-3 model. Packet size is fixed as 20 Byte
UE distribution	20% of users are outdoors (3 km/h) 80% of users are indoor (3 km/h)
BS receiver	CBC MAP detector
UL power control	Open loop power control $\alpha = 1$, $P_0 = -95$ dBm
Channel estimation	Ideal channel estimation
Retransmission Timer	1–10 s

single user detector is applied in OFDMA, users that collide with each other are not considered to transmit successfully.

Figure 3.46 gives the IGMA simulation results. As can be observed from the figure, for given packet arrival rate, the packet loss rate of IGMA is lower than that of OFDMA. For 10% packet loss rate, the supported packet arrival time is more than 5 times larger than that of OFDMA. This is a huge improvement compared with OMA scheme and the connection density can be also improved with the aid of IGMA.

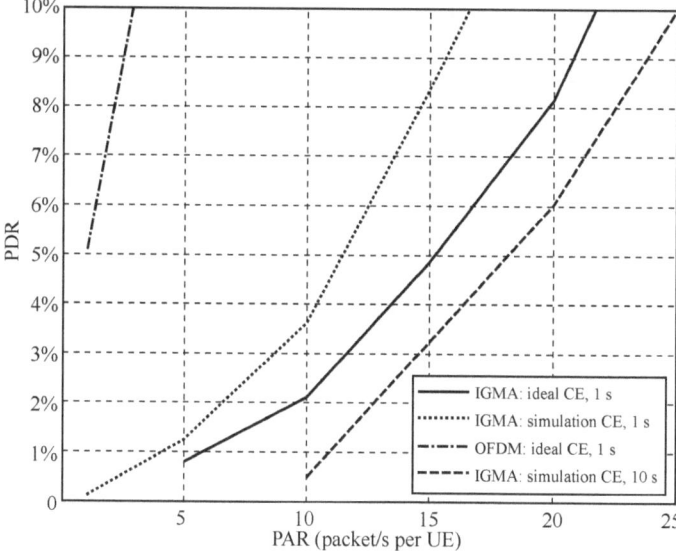

Figure 3.46: Evaluation results for system packet loss rate of different schemes.

From Figure 3.46, it can be observed that for 1% packet loss rate, the achievable packet arrival of IGMA with 10s retransmission timer is more than two times higher than that of IGMA with ideal channel estimation. Based on above evaluation results and the requirements on packet arrival time for mMTC scenario, the corresponding connection density can be calculated. Take IGMA with ideal channel estimation and 1s dropping timer as an example. Assume that the average packet arrival interval per UE for mMTC scenario $\lambda_{mMTC} = 7,680$ s, the packet arrival rate is 5.82 packets/s, as a result the connection density (the bandwidth is 6 PRBs) can be expressed as follows:

$$\rho_{mMTC} = \frac{N_{UE}\lambda_{SLS}}{\lambda_{mMTC}} \frac{1}{S_{sec\,tor}} = \frac{20 \times 7680 \times 5.82}{S_{sec\,tor}} = 1.03 \times 10^6 \qquad (3.42)$$

Table 3.17 gives corresponding connection density of different target packet arrival time in mMTC scenario. From Table 3.17, it can be seen that IGMA can fulfil the requirements on connection density for mMTC scenario.

Table 3.17: Connection density of different target packet arrival time.

Ratio factor based on NB-IOT	Target packet arrival time/h	Connection density/$10^6 \cdot km^{-2}$
1	10.575	21.1
1/2	5.2875	10.6
1/5	2.115	4.22
1/10	1.0575	2.11
1/20	0.5287	1.06

3.7 Other Candidate Multiple Access Technologies

Jie Zeng

3.7.1 BDM

Traditional linear superposition code implements linear division directly to power resource of discrete symbols and allocates fixed power resource to each user. Constellation mapping is independent among users. BDM can be considered as an extension to traditional hierarchical modulation technology. The nature of BDM is a non-linear superposition coding technology. Different from traditional linear superposition code, BDM regards power resource of multiple discrete symbols as a whole. High-order constellation symbols are transmitted directly, meanwhile bit resources carried by high-order constellation symbols are divided. It is essentially non-linear division to power resource of discrete symbols, without independent constellation mapping of each user. Take four users as an example, BDM channel resource division is illustrated in Figure 3.47. Bits in different shadows are allocated to different users.

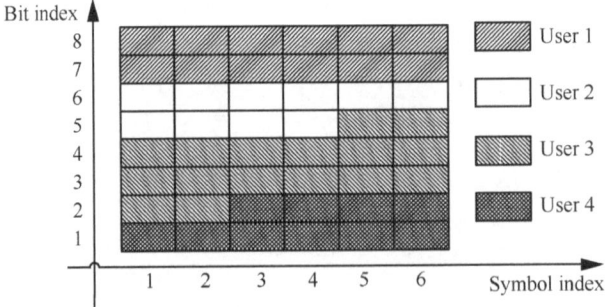

Figure 3.47: BDM channel resource division diagram.

BDM transmitter in discrete time baseband model is illustrated in Figure 3.48. Channel coding and bit interleaving are carried out independently from user (group) 1 to user (group) K in order to obtain interleaved bits of different users (groups); Referring to the channel resource division pattern illustrated in Figure 3.47, interleaved bits from user (group) 1 to user (group) K are combined in the bit aggregation module to get overall interleaved bits; After high-order constellation mapping of overall interleaved bits, symbols to be transmitted are obtained.

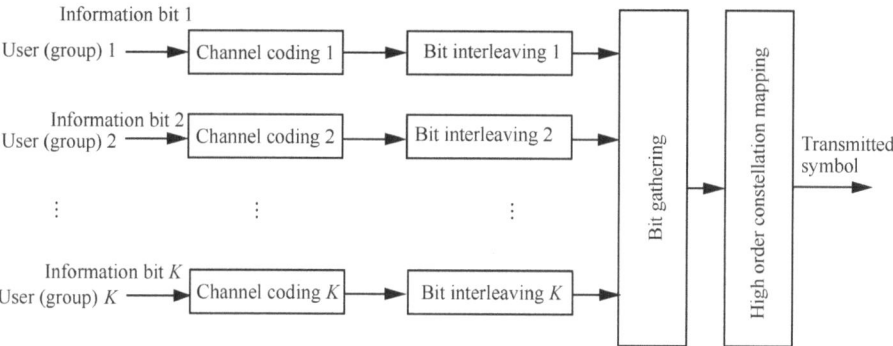

Figure 3.48: BDM transmitter.

One kind of BDM receiver using Single Stage Decoding (SSD) is illustrated in Figure 3.49. With the help of CSI, constellation de-mapping module implements constellation de-mapping to received symbols in order to get soft information of overall interleaved bits. Referring to the channel resource division pattern illustrated in Figure 3.47, soft information of overall interleaved bits are decomposed in bit decomposition module to get interleaved bit soft information of user i. Interleaved bit soft information of user i is then handled by bit soft information de-interleave module and channel decoding module to obtain information bit soft information or decision results of user i.

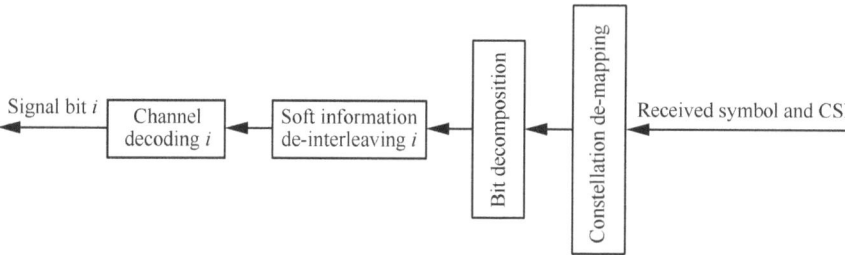

Figure 3.49: Block diagram of BDM receiver using SSD (user i).

Multi-Stage Decoding (MSD) can further improve the receiving performance of BDM scheme. Without loss of generality, suppose the priorities decrease in turn from user 1 to user K (usually high priority means low decoding SNR threshold at the receiver). Take two users as an example, the receiver of user 1 coincides with the receiver with SSD. Decoded bit soft information or decision result of user 1 is feedback to high-order constellation mapping module to assist decoding of user 2.

Preliminary research shows that when the complexity is the same, BDM can be closer to the upper bound of multi-user joint transmission rate compared with simple linear composition coding. Besides, BDM can apply Bit Interleaved Code Modulation (BICM) system and SSD receiving algorithm design, which has low complexity and small performance loss compared with MSD [41].

3.7.2 IDMA

IDMA distinguishes different users by using different chip-level interleavers. In essence, IDMA is still a spread spectrum technology implemented by coding; therefore, it inherits the advantage of Spread Spectrum Communication and CDMA. Meanwhile IDMA can achieve good multi-user detection performance and low complexity at the receiver. In IDMA systems, interleaver is the multiple access identification of different users, hence the orthogonality quality of interleaver group, which constitute by interleavers of all users. It determines the system detection performance and the ability of resisting multiple access interference and noise.

The basic principle of IDMA is that different interleaving modes are used by different users, and interleaver is used to distinguish users. Interleaver is located after coding and spreading. Its function is to make bits which are connected with codeword sequence correlated with each other. Therefore, the bandwidth occupied by spreading code in traditional CDMA systems can be released and used in channel coding. Coding and spreading are completed before interleaving, and spreading sequence is the same to all users. Therefore, interleaving in IDMA systems belongs to chip-level interleaving.

Figure 3.50 shows the transmitter and receiver structure of an IDMA system with K users [42]. The input data sequence d_k of user k ($k \leq K$) is encoded based on a low-rate encoder, resulting in a coded sequence $\mathbf{c}_k = [c_k(1), \cdots, c_k(j), \cdots, c_k(J)]^T$ with frame length J. The element in $\mathbf{c}_k$ is the coded bit. The coded sequence is then interleaved by a chip-level interleaver π_k, producing $\mathbf{x}_k = [x_k(1), \cdots, x_k(j), \cdots, x_k(J)]^T$. From the receiving diagram, it can be seen that users are distinguished only through different interleavers, thus IDMA is an interleaving multiple access technology.

The receiver of IDMA consists of two parts, which are ESE and K posteriori probability decoders. Received signal $\mathbf{r}$ is sent to ESE first, then multi-user detection is implemented by information exchange and iterative decoding between ESE and K posteriori probability decoders. Each decoder in the K posteriori probability decoders corresponds

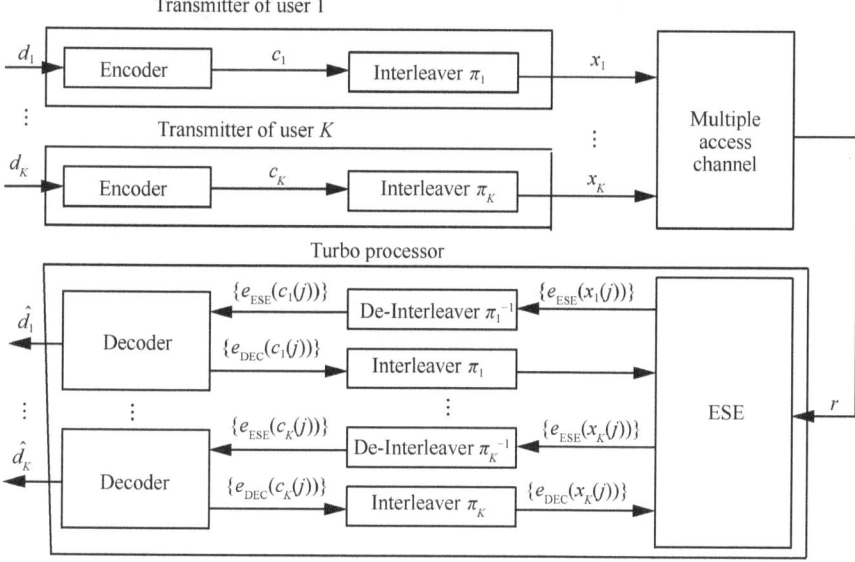

Figure 3.50: Transmitter and receiver structure of an IDMA system.

to one user [42]. Iteration does not need matrix operations. ESE has a low complexity, and operation complexity of user chip is independent with user number. When there are a large number of access users, the effect of IDMA iterative receiving algorithm to resist multiple access interference is obvious [43].

3.7.3 LSSA

In the 3GPP RAN WG1 #85 meeting, Electronics and Telecommunications Research Institute (ETRI) suggested that LSSA should be considered as one of the candidate non-orthogonal multiple access schemes. LSSA supports competitive transmission and grant-free access.

The low-rate FEC encoder in LLSA can reduce multi-user interference effectively. In LSSA, each user data is bit-level or symbol-level and has multiplexed with certain signature sequence pattern. The multiplexed signature sequence pattern is not known to other users. Figure 3.51 shows the LSSA transmitter structure in uplink mMTC scenario.

Information of each user is encoded by low-rate channel coding, which can be replaced by high-rate channel coding module with non-orthogonal spreading sequences, then coded user signal is bit-level or symbol-level signal that has already multiplexed with signature sequence vector.

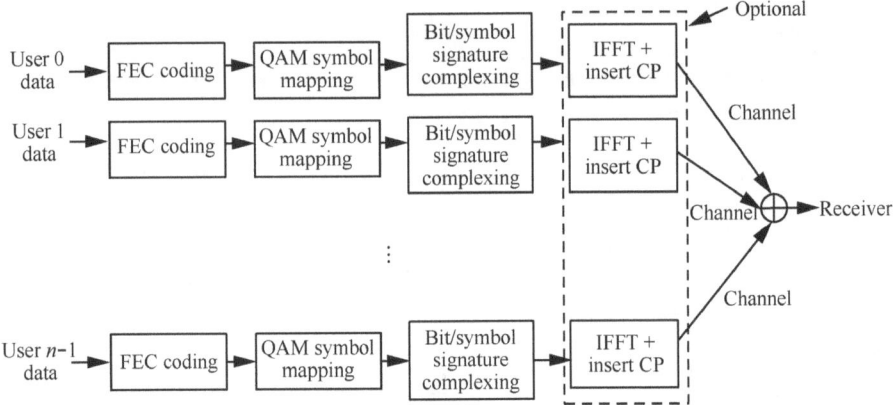

Figure 3.51: LSSA transmitter structure.

User signature sequence is composed of reference signal, binary complex sequence and transpose vector. Long user signature sequence is appropriate for competitive transmission in uplink mMTC scenario, in which low cross-correlation among different user signature sequences is needed. However, long user signature sequence vector will bring high decoding complexity and higher latency, thus the length of user signature sequence vector should be proper. Even with short user signature sequence vector, the multiplexed user number in uplink is not limited by the signature sequence vector length. Signal detection at the receiver does not rely on orthogonal multiplexing code, therefore user overload can be supported.

All the signature sequences assigned to users have the same length and should be random chosen by mobile terminal or directly assigned by network. LSSA can realize low latency by fully utilizing frequency diversity.

LSSA supports uplink asynchronous transmission multiplexing. Even the timing of one user is different from other users; user signals that are superposed on same resources can be detected by utilizing correlation among different signature sequence patterns. However, due to the high complexity at the BS, timing difference among users is limited by pre-defined time value, which is the data round trip time between cell-edge users and cell-center users [44].

3.7.4 NOCA

Similar with other non-orthogonal multiple access technologies based on spreading, the basic idea of NOCA is spreading data with non-orthogonal sequences before data is transmitted. Spreading can be configured in frequency, time domain.

Basic structure of NOCA transmitter is illustrated in Figure 3.52. In this scheme, frequency spreading based on OFDM waveform is used, SF is the spreading factor,

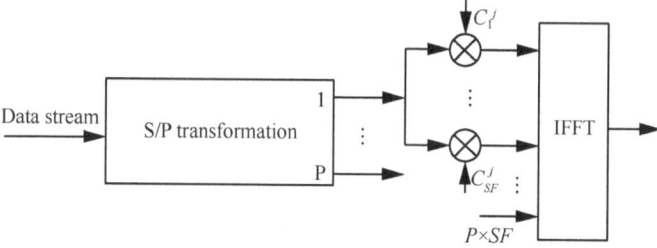

Figure 3.52: NOCA transmitter structure.

C_s^j denotes the spreading sequence of user j, where s is the subcarrier number occupied by user j. Initial modulated signal is transformed to P parallel sequences by a serial to parallel converter, and each sequence is mapped to SF subcarriers. Therefore $P \times SF$ subcarriers are needed to transmit data. By using different spreading sequences, multiple users can multiplex in given time-frequency resources [41, 45].

Spreading sequences used in NOCA should have characteristics such as good self-correlation, good cross-correlation, short memory, low complexity and so on [45].

Massive non-orthogonal sequences are used in NOCA, which could decrease collisions of competitive access in code domain [41]. Preamble and spreading data are transmitted together in NOCA channel based on competitive access. In order to increase preamble detection probability, orthogonal sequence can be used as preamble. Each preamble P_j corresponds to a spreading code C_j. If preamble is detected, spreading code corresponding to the preamble is obtained by the receiver. Since the applied non-orthogonal spreading code has low cross-correlation, when the concurrent user number is small, data can be handled by simple linear receivers. When the concurrent user number is large, more advanced receivers are needed for data processing. In NOCA channel based on competitive access, each data group composed by P_j and C_j is considered as one transmission opportunity. Transmission opportunity number is equal to the minimum value of NOCA codebook length and the number of preambles. Applying orthogonal sequence longer than SF can generate more preambles easily, thus transmission opportunity number is only limited by NOCA codebook length [46].

3.7.5 NCMA

NCMA is a multiple access scheme that implements spreading with non-orthogonal codes, whose codewords are obtained by the line packet problem of Glassman manifold.

Codebook is defined as:

$$\mathbf{C} = \left[\mathbf{c}^{(1)} \cdots \mathbf{c}^{(K)} \right] = \begin{bmatrix} c_1^{(1)} & \cdots & c_1^{(K)} \\ \vdots & \ddots & \vdots \\ c_N^{(1)} & \cdots & c_N^{(K)} \end{bmatrix}, \quad \mathbf{C} \subset C^{N \times K} \tag{3.43}$$

where N is the spreading factor, K is the superposition factor. Codebook design can be expressed as maximization of the minimum chordal distance of code pairs, which is:

$$\min \left(\max_{1 \le k < j \le K} \sqrt{1 - \left| \mathbf{c}^{(k)^*} \mathbf{c}^{(j)} \right|^2} \right) \tag{3.44}$$

where $\mathbf{c}^{(k)^*}$ is the conjugate code of $\mathbf{c}^{(k)}$. Non-orthogonal code obtained by the line packet problem of Glassman manifold provides following two conditions.

1) Correlation of code pairs:

$$\begin{cases} \left| \mathbf{c}^{(k)^*} \mathbf{c}^{(j)} \right| = 1, \forall k, k = 1, \cdots, K \\ \text{if } N > K, \left| \mathbf{c}^{(k)^*} \mathbf{c}^{(j)} \right| = \delta_{N,K}, \forall k, \forall j, k = 1, \cdots, K, j = 1, \cdots, K \\ \text{if } N \le K, \left| \mathbf{c}^{(k)^*} \mathbf{c}^{(j)} \right| = 0, \forall k, \forall j, k = 1, \cdots, K, j = 1, \cdots, K \end{cases}$$

2) Lower bound of cross-correlation of code pairs: $\delta_{N,K} \ge \sqrt{1 - \dfrac{(N-1)K}{N(K-1)}}$

Upper bound of chordal distance of code pairs: $\sqrt{1 - \left| \mathbf{c}^{(k)^*} \mathbf{c}^{(j)} \right|^2} \le \dfrac{(N-1)K}{N(K-1)}$

The receiver structure of NCMA is illustrated in Figure 3.53, in which Non-Orthogonal Code Cover (NCC) allocates one non-orthogonal code to each user.

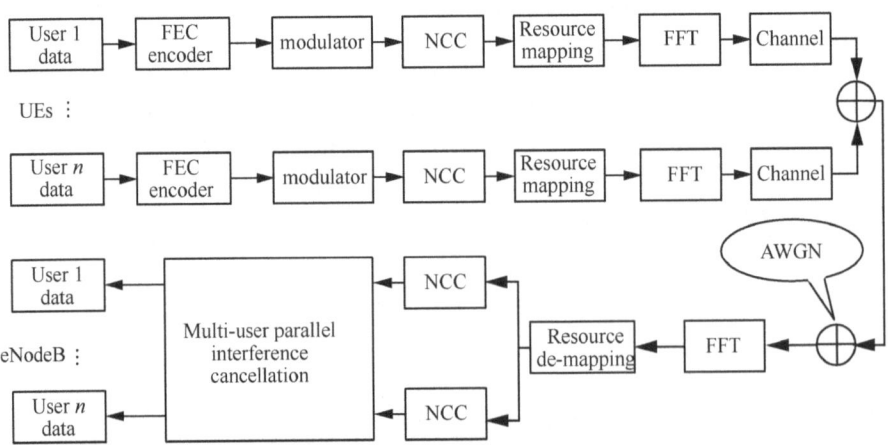

Figure 3.53: NCMA receiver structure in uplink.

NCMA can achieve a higher system throughput with small BLER. In addition, NCMA can improve the connection performance and satisfy the QoS requirement. The receiver of NCMA employs PIC, which could realize multi-user detection with low complexity. In mMTC scenario, NCMA can support massive small packet connections without changing the transport block size [47].

3.7.6 Summary

Besides the above multiple access technologies, there are some other multiple access technologies including FDS [48], LDS-SVE [49], Low Code Rate Spreading (LCRS) [48], RDMA [50], GOCA [50] and so on. Since these technologies all apply non-orthogonal multiple access, the principles are much the same, and will not be repeated in this section. Compared with OMA, all of these non-orthogonal multiple access schemes realize the goal of multiplexing data of multiple users on one resource and achieve a higher system performance gain. They will be candidate technologies in future NMA schemes. Predictably, their technologies will also play an important role in the future 5G standardization process at home and abroad.

References

[1]   IMT2020_TECH_N_140015_SCMA 技术介绍_华为[Z]. 2015.
[2]   Whitepaper v2.0D. Alternative multiple access v1[Z]. Future Forum, 2015.
[3]   R1-162155. Sparse code multiple access (SCMA) for 5G radio transmission[R]. Huawei, HiSilicon, 3GPP TSG RAN WG1 Meeting #84bis, Busan, Korea, 2016.
[4]   Lu L, Chen Y, Guo W T, et al. Prototype for 5G New air interface technology SCMA and performance evaluation[J]. China Communications, 2015, 12(s1): 38–48.
[5]   R1-162226. Discussion on multiple access for new radio interface[R]. 3GPP TSG RAN WG1 Meeting #84bis, Busan, Korea, 2016.
[6]   R1-1608855. Advanced multi-user detectors for grant-free transmissions[R]. Huawei, HiSilicon 3GPP TSG RAN WG1 Meeting #86bis, Lisbon, Portugal, 2016.
[7]   Mu H, Ma Z, Alhaji M, et al. A fixed low complexity message pass detector for up-link SCMA system[J]. IEEE Wireless Communications Letters, 2015, 4(6): 585–588.
[8]   Xiao K, Xiao B, Zhang S, et al. Simplified multiuser detection for SCMA with sum-product algorithm[C]. WCSP, 2015: 1–5.
[9]   R1-1608853. Further UL LLS results[R]. Huawei, HiSilicon, 3GPP TSG RAN WG1 Meeting #86bis, Lisbon, Portugal, 2016.
[10] R1-167335. LLS results for DL MA schemes[R]. Huawei, HiSilicon, 3GPP TSG RAN WG1 Meeting#86, Gothenburg, Sweden, 2016.
[11] R1-167336. SLS results for SCMA in DL eMBB scenario[R]. Huawei, HiSilicon, 3GPP TSG RAN WG1 Meeting #86, Gothenburg, Sweden, 2016.
[12] R1-166097. SLS results for MA evaluation in mMTC scenario[R]. Huawei, HiSilicon, 3GPP TSG RAN WG1 Meeting #86, Gothenburg, Sweden, 2016.

[13] Chen S, Ren B, Gao Q, et al. Pattern division multiple access (PDMA)-a novel non-orthogonal multiple access for 5G radio networks[J]. IEEE Transactions on Vehicular Technology, 2017, 66(4): 3185–3196.

[14] R1-163383. Candidate solution for NMA_final[R]. CATT, 3GPP TSG RAN WG1 Meeting #84bis, Busan, Korea, 2016.

[15] R1-166469. Update of LLS results of PDMA[R]. CATT, 3GPP TSG RAN WG1 Meeting #86, Gothenburg, Sweden, 2016.

[16] R1-167870. Update of LLS results of PDMA[R]. CATT, Gothenburg, Sweden, 2016.

[17] R1-166470. Initial SLS results of PDMA[R]. CATT, 3GPP TSG RAN WG1 Meeting #86, Gothenburg, Sweden, 2016.

[18] R1-166468. Remaining issues on evaluation assumption and methodology of MA[R]. CATT, 3GPP TSG RAN WG1 Meeting #86, Gothenburg, Sweden, 2016.

[19] 袁志峰, 郁光辉, 李卫敏. 面向5G 的MUSA 多用户共享接入 [J]. 电信网技术, 2015, (5): 28–31.

[20] R1-162226. Discussion on multiple access for new radio interface[R]. ZTE, 3GPP TSG RAN WG1 Meeting #84bis, Busan, Korea, 2016.

[21] IMT2020_TECH_NMA_140030. 非正交多址复用及MUSA (Multi User Shared Access)[Z]. ZTE, 2014.

[22] R1-164270. Receiver implementation for MUSA[R]. ZTE, 3GPP TSG RAN WG1 Meeting #85, Nanjing, China, 2016.

[23] R1-166404. Receiver details and link performance for MUSA[R]. ZTE Microelectronics, 3GPP TSG RAN WG1 Meeting #86 Gothenburg, Sweden, 2016.

[24] R1-166402. Remaining issues on multiple access evaluation assumption[R]. ZTE Microelectronics,3GPP TSG RAN WG1 Meeting #86, Gothenburg, Sweden, 2016.

[25] Yan C, Harada A, Benjebbour A, et al. Receiver design for downlink non-orthogonal multiple access (NOMA)[C]. IEEE VTC Spring, 2015: 1–6.

[26] R1-163111. Initial views and evaluation results on non-orthogonal multiple access for NR uplink[R]. NTT DOCOMO, INC., 3GPP TSG RAN WG1 Meeting #84bis, Busan, Korea, 2016.

[27] Li A, Lan Y, Chen X, et al. Non-orthogonal multiple access (NOMA) for future downlink radio access of 5G[J]. China communications, 2015, 12(Supplement): 28–37.

[28] Kountouris M, Gesbert D. Memory-based opportunistic multi-user beamforming[C]. International Symposium on Information Theory, ISIT 2005, 2005: 1426–1430.

[29] R1-163510. Candidate NR Multiple Access Schemes[R]. Korea Qualcomm Incorporated, 3GPP TSG-RAN WG1 #84b, Busan, 2016.

[30] R1-164689. RSMA and SCMA comparison[R]. Qualcomm Incorporated, 3GPP TSG-RAN WG1 #85, Nanjing, China Qualcomm Incorporated, 2016.

[31] R1-166358. RSMA and SCMA comparison[R]. Qualcomm Incorporated, 3GPP TSG-RAN WG1 #86, Gothenburg, Sweden, 2016.

[32] R1-1610118. Link level simulation results for RSMA vs OMA[R]. Qualcomm Incorporated, 3GPP TSG-RAN WG1 #86bis, Lisbon, Portugal Lisbon, Portugal, 2016.

[33] R1-1610119. System level simulation results for calibration[R]. Qualcomm Incorporated, 3GPP TSG-RAN WG1 #86bis, Lisbon, Portugal, 2016.

[34] R1-1610120. System level simulation results for proposed multiple access scheme[R]. Qualcomm Incorporated, 3GPP TSG-RAN WG1 #86bis, Lisbon, Portugal, 2016.

[35] R1-163992. Non-orthogonal multiple access candidate for NR[R]. 3GPP TSG-RAN WG1 Meeting #85, Nanjing, China, Samsung, 2016.

[36] IMT 2020_TECH_NMA_16039. Inter leaver-grid multiple access[Z]. Samsung, 2016.

[37] R1-166750. Link level performance evaluation for IGMA[R]. Samsung, 3GPP TSG RAN WG1 Meeting #86, Gothenburg, Sweden, 2016.

[38] R1-166751. System level performance evaluation for nonorthogonal multiple access[R]. Samsung, 3GPP TSG RAN WG1 Meeting #86, Gothenburg, Sweden, 2016.

[39] 3GPP TR45.820. Cellular system support for ultra-low complexity and low throughput Internet of Things (CIoT)[S]. 2015.

[40] R1-1609042. System level evaluation results for IGMA[R]. Samsung, 3GPP TSG RAN WG1 Meeting #86bis, Lisbon, Portugal, 2016.

[41] Jin H, Peng K, Song J. Bit division multiplexing for broadcasting[J]. IEEE transactions on broadcasting, 2013, 59(3): 539–547.

[42] Li P, Liu L H, Wu K Y, et al. Interleave division multipleaccess[J]. IEEE transactions on wireless communications, 2006, 5(4): 938–947.

[43] Mahafeno I M, Langlais C, Jego C. CTH12-4: Reduced complexity iterative multi-user detector for IDMA (interleavedivision multiple access) system[C]. IEEE GLOBECOM 2006, San Francisco, CA, 2006: 1–5.

[44] R1-164869. Low code rate and signature based multiple access scheme for New Radio[R]. ETRI, 3GPP TSG RAN WG1 Meeting #86, Gothenburg, Sweden, 2016.

[45] R1-165019. Non-orthogonal multiple access for new radio[R]. Nokia, 3GPP TSG RAN WG1 Meeting #85, Alcatel-Lucent Shanghai Bell, Nanjing, China, 2016.

[46] R1-167249. Non-orthogonal coded access (NOCA)[R]. Nokia, Alcatel-Lucent Shanghai Bell, 3GPP TSG RAN WG1 Meeting #86, Gothenburg, Sweden, 2016.

[47] R1-162517. Considerations on DL/UL multiple access for NR[R]. LG Electronics, 3GPP TSG RAN WG1 Meeting #84bis, Busan, Korea, 2016.

[48] R1-162385. Multiple access schemes for new radio interface[R]. Intel Corporation, 3GPP TSG RAN WG1 Meeting #84bis, Busan, South Korea, 2016.

[49] R1-164329. Initial LLS results for UL non-orthogonal multiple access[R]. Fujitsu, 3GPP TSG RAN WG1 Meeting #85, Nanjing, China, 2016.

[50] R1-167535. New uplink non-orthogonal multiple access schemes for NR[R]. MediaTek Inc., 3GPP TSG RAN WG1 Meeting #86, Gothenburg, Sweden, 2016.

Chapter 4
Application of NMA Technologies in 5G

NMA technology can satisfy different requirements of multiple future 5G scenarios (including eMBB, URLLC and mMTC).

The main requirements of the eMBB scenario are: higher system capacity, higher data rate and higher spectrum efficiency. In this scenario, target peak throughput of data service in downlink is 20 Gbit/s, while target peak throughput in uplink is 10 Gbit/s; downlink spectrum efficiency is 30 bit/s·Hz^{-1}, while uplink spectrum efficiency is 15 bit/s·Hz^{-1}. In LTE, in order to realize higher spectrum efficiency, scheduling and control mechanism is employed, which uses a large amount of signaling. In eMBB downlink, NMA technology supports multiple data stream transmission on each time-frequency resource, which can improve system capacity and spectrum efficiency of cell-edge users. In eMBB uplink, like orthogonal multiple access in LTE, NMA technology also needs scheduling and control mechanism.

The main requirements of URLLC are low latency and high reliability. Reliability can be evaluated by transmission success rate of X bytes of data in 1 ms, which is the transmission time of a small packet from wireless protocol layer L2/L3 of the transmitter to the wireless protocol layer L2/L3 of the receiver with certain channel quality. In this scenario, the target of reliability is achieving a 99.999% transmission success rate in 1 ms. In URLLC, latency is related to error probability and the corresponding HARQ. Target latency of user plane in both uplink and downlink is 0.5 ms. Interference cancellation is crucial for an ultrareliable system. Orthogonal multiple access schemes use orthogonal sub-channel for interference-free data transmission; NMA schemes can transmit more data with same resource within one-unit time to improve system capacity, however, interference exists in this situation, thus advanced interference cancellation technologies are needed at the receiver to ensure high reliability of information.

In mMTC scenario, machine-to-machine communication occupies a large amount of data, whose total connection number will reach 100 billion. For uplink transmission, the core requirement of mMTC is to accommodate a large number of occasional small packet services, and each small packet service has the characteristics of low-cost and high-energy efficiency, which is conducive to large-scale deployment. In mMTC scenario, target connection density of access equipment in urban environment reaches 10^6/km^2, and terminal battery lifespan reaches 15 years [1, 2].

https://doi.org/10.1515/9783110666366-004

4.1 Support for eMBB Scenario

Bin Ren

4.1.1 eMBB Scenario Overview

The target of eMBB is to further improve the performance and user experience based on current MBB. Its scenario can be divided into low frequency and high frequency. Low-frequency resource below 6 GHz is much important to coverage enhancement for high capacity and high rate speed scenario in eMBB; High frequency and broad bandwidth are effective means to improve system capacity in hotspots. High- and low-frequency cooperation is the basic method to satisfy the requirements in eMBB. These two cases correspond to two applications in eMBB, which are wide area coverage and local hotspot, respectively, as illustrated in Figure 4.1. Local hotspot needs to support not only a higher user density but also a higher user throughput in hot areas. Compared with wide area coverage, although the requirements of user mobility is low in local hotspot, the user transmission rate is high. Wide area coverage needs to implement seamless coverage of wireless networks and satisfy user experience when the moving speed is high. However, the requirements of transmission rate are lower than that of local hotspot [3].

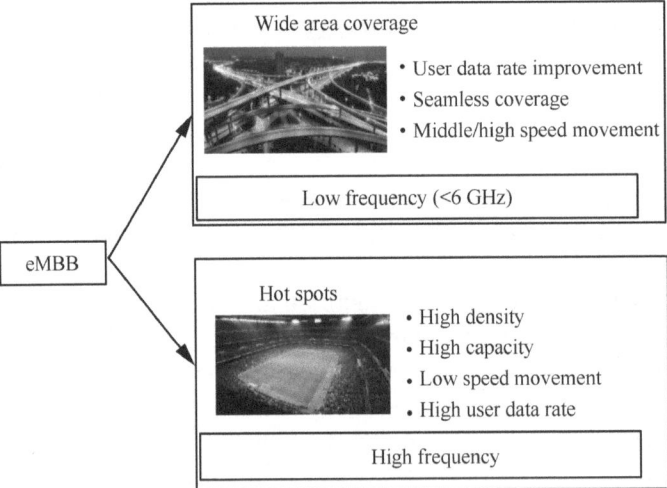

Figure 4.1: Application of eMBB scenario.

Reference [4] gives application prospects of NMA technology in eMBB scenario.

High network capacity: compared with orthogonal multiple access in LTE, NOMA requires less orthogonality. In addition, NOMA can transmit more data with same time-frequency resource in both uplink and downlink.

High user density: inter-user interference can be handled efficiently, which means higher user density and higher traffic load can be achieved.

Unified user experience: cell-edge users and cell-center users have unified user experience and high user speed can be supported.

Mixed service type transmission: transmission of different service types is efficiently balanced, which makes data transmission more efficient. For example, balanced transmission of big packet video service with large amount of data and small packet service with low latency requirement.

Application examples of eMBB include video stream, video phone and so on. For enhanced multimedia service, large packet transmission mainly appears in downlink, while some small packet transmission is mainly used for information interaction or video control in uplink [5].

4.1.2 Applications of NMA in eMBB Scenario

The target of new wireless access systems is to support wide area coverage, indoor and hotspots scenarios. Hotspots include outdoor court, indoor stadium, shopping mall, auditorium, outdoor square and so on. There are a large amount of users with high rate requirement in such areas. Due to the limitation of time/frequency/space resource, it is difficult for OMA to support so many users with high rate. In the 3GPP RAN WG1 #84bis meeting, the participants all agreed that NMA schemes can be used in diversified application scenarios and cases in NR. One of the most important requirements in eMBB is to realize higher capacity gain in both uplink and downlink and higher spectrum efficiency. To satisfy the requirements including massive connection, large capacity, low latency in 5G applications, NMA technology is urgently needed, and NOMA technology becomes an important candidate technology [6].

Suitable NOMA schemes are needed in eMBB. Reference [7] summarizes NOMA schemes that can be used in eMBB: SCMA, PDMA, NOMA, MUSA and BDM. Among these technologies, MUSA and BDM are mainly used in eMBB uplink.

SCMA has the advantages of high capacity, massive connections, low overhead and so on. The code domain sparsity in SCMA can decrease interference among layers, collision probability of transmitted symbols and reduce receiver complexity. Besides, since SCMA employs reduced order codebooks with reduced constellations, the projection point number in codebook is decreased, which makes multidimensional modulation transmission more efficient, and the complexity is lower. In single cell and multi-cell scenarios, SCMA can provide open loop multi-user multiplexing, unified user experience and improve system capacity. In addition, SCMA has strong robustness to interference [8]; thus, it is suitable for eMBB scenario.

PDMA uses time domain, frequency domain, space domain and power domain resource or combination of the above resource. Compared with OMA, PDMA can obtain larger channel capacity and higher spectrum efficiency; therefore, it is more suitable for eMBB. Reference [6] describes the deployment of PDMA system with intensive uplink users in eMBB scenario, such as World Cup with dense population, which has high requirements for the access capacity. Compared with OMA, PDMA can realize multiplexed transmissions of more users on same wireless resource and connections of high density users, which benefits more people [6].

By superposing signals in power domain, same time-frequency resource can be shared by multiple users in NOMA, system capacity and throughput are improved. Figure 4.2 shows the capacity region comparison of OMA and NOMA in uplink when SNR difference among user pairs achieves 20 dB, where capacity gain is illustrated by the shaded area. From the figure, it can be seen that compared with OMA, NOMA improves the system capacity region, thus it can satisfy the large capacity requirement in eMBB. Besides, NOMA does not rely on CSI of frequency selective fading. Thereby UE speed and CSI feedback latency will not affect system performance in actual wide area coverage applications, which means NOMA system has strong robustness [3, 9]. The above characteristics of NOMA meet the needs of eMBB scenario.

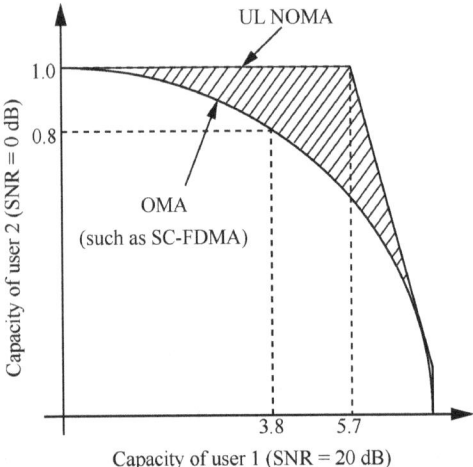

Figure 4.2: Capacity region comparison between OMA and NOMA.

MUSA allocates different code sequences to different users, which is equivalent to spreading in essence, such as spread each bit into 3 bits. MUSA code sequence is actually a short sequence multivariate code with low correlation, complex field and constellation. When user channel conditions are different, code sequence can be determined in a relative relaxed environment, which guarantees high system capacity and fairness among users. MUSA can support a high reliable access number which is several times the access number of OMA in same time-frequency resource, thus can simplify resource scheduling and reduce access time in massive access. MUSA has low implementation difficulty and controllable system complexity, can support a large number of access users, does not need synchronization in principle and can improve the terminal battery life, all of which are suitable for the application of IoT.

In NOMA, competitive-based grant-free schemes can improve connection efficiency, reduce signaling overhead and decrease access latency, which are not only applicable for mMTC, URLLC, but also for eMBB [5]. Grant-free transmission schemes are conducive to uplink small packet data transmission, while grant-free transmissions of large packet data will bring high system complexity. A mix of non-orthogonal multiple access and OMA can solve this problem. Reference [10] proposes a grant-free-based uplink transmission scheme in eMBB by combining OMA and NOMA. The following is a transmission scheme in uplink by combining OMA and NOMA.

1) UE is authorized to transmit data with grant-free NOMA schemes;
2) UE transmits small packet data, such as TCP, ACK and so on, by using NOMA resource blocks;
3) Once UE has large packet data to transmit, UE sends the buffer status to BS;
4) BS assigns OMA resource to designated UE, such as PRB group and so on;
5) UE transmits large packet data on OMA resource.

4.1.3 Performance Evaluation of NMA in eMBB Scenario

Downlink system-level simulation of SCMA in eMBB scenario is carried out in reference [11]. In the simulation there are two business models: full buffer and non-full buffer. In full buffer model, open loop MIMO is used, and SCMA throughput gain compared with OFDMA is outputted; while in non-full buffer model, the influence of whitening interference is emphasized, where whitening interference is produced in downlink small packet data transmission of SCMA.

In full buffer transmissions, MU-SCMA employs algorithms based on code domain pairing or power domain pairing, which is the same as the MUST scheme. The goal is to maximize weighted overall rate. For each RBG, two UEs with maximum weighted overall rate can be paired. UE close to BS eliminates the interference from UE far away from BS by employing ideal codeword-level interference cancellation receive, and obtains the needed information by further decoding; UE far away from BS processes the signal of user close to base station as interference. In non-full buffer transmissions, outdated measurement information cannot be used, because some TP are not always transmitting on same frequency resources, thus real time feedback measurement results are needed to improve accuracy of feedback. Reference [11] gives simulation analysis to full buffer and non-full buffer schemes in dense urban area. In non-full buffer situations, the system bandwidth is 10 MHz, and data packet is 6 Kbit. Some other specific simulation parameters are referred to reference [11].

Figure 4.3 shows the performance gain comparison between MU-SCMA and OFDMA in full buffer situations [11]. From Figure 4.3, it can be seen that SCMA

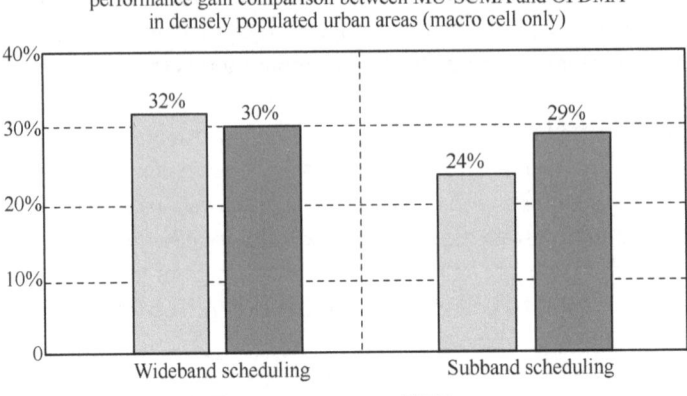

Figure 4.3: Performance gain of MU-SCMA and OFDMA in full buffer situations.

has more obvious gain than OFDMA in full buffer situations. From Figure 4.4, it can be seen that SCMA has more obvious User Perceived Throughput (UPT) gain than OFDMA in non-full buffer situations.

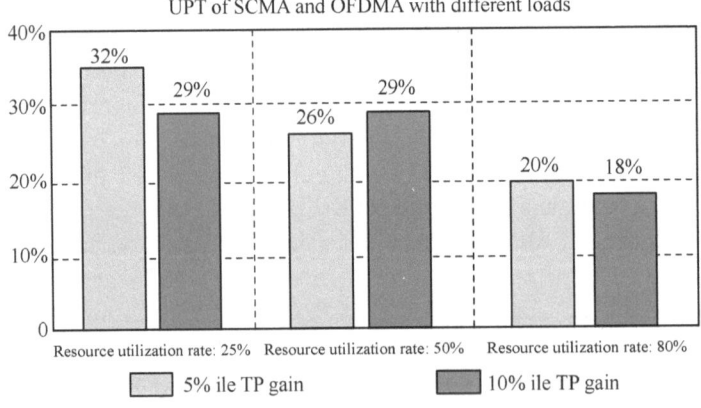

Figure 4.4: UPT of SCMA and OFDMA with different loads.

Reference [12] gives preliminary evaluation to uplink NOMA performance in eMBB scenario. From the performance comparison between NOMA and SC-FDMA, it can be seen that NOMA has a better uplink cell throughput than SC-FDMA. Based on this, FFR can coordinate inter-cell interference better and achieve better performance.

Reference [13] gives spectrum efficiency evaluation results when NOMA is used in eMBB scenario. With the help of NOMA, uplink spectrum efficiency improves more than 30% compared with OMA, while downlink spectrum efficiency improves mare than 100%. The huge gain of NOMA in spectral efficiency indicates that it can be applied to eMBB scenario.

4.2 Support for URLLC Scenario

Jie Zeng

4.2.1 URLLC Scenario Overview

As one of the important application scenarios in 5G, URLLC has high requirements for system throughput, delay and reliability. In URLLC, radio transmission latency, network forwarding latency and retransmission probability should be decreased as much as possible to meet the requirements of ultrareliable and low latency. From the perspective of reducing latency, short frame structure applied by the system can help to meet the transmission latency requirement of less than 1 ms with one retransmission; Meanwhile, in order to achieve 99.999% system reliability, multiple antennas can be deployed at the terminal to increase diversity gain. The application cases of URLLC include wireless control of industrial manufacturing, remote medical surgery, transportation safety and so on [1, 14].

In URLLC, different application cases have different requirements. Reference [15] lists specific requirements of some typical application cases, as given in Table 4.1. By introducing technologies including grant-free NMA in URLLC signaling overhead and data relay can be reduced. If more advanced modulation coding and MIMO technology are adopted, transmission reliability can be further improved.

Table 4.1: Typical URLLC use cases and requirement.

Typical use case	Deployment scenario	Service characteristic	Latency requirement	Reliability requirement
eV2X	Dense urban / Uma / Rma	CAM/DENM/security intelligence	<10 ms	<10e-5
Augmented Reality	Indoor / Dense urban / Uma / Rma	8 K stereo video stream	<10–20 ms	/
Extreme Industrial Control	Indoor hotspot	High-fidelity control and interaction, periodic and event-triggered, small/ medium packet	<1–10 ms	<10e-9
eHealth	Deep Indoor	High-fidelity control and interaction, periodic and event-triggered, small/ medium packet	<1–10 ms	<10e-5
Smart Grid	Uma/Rma	Monitor and dynamic power control	<1 ms	<10e-5

4.2.2 Applications of NMA in URLLC Scenario

Two main performance indexes of URLLC are latency and reliability. 5G NR supports ultrareliable and low latency scenario, which needs to consider frame structure, HARQ, uplink access, channel coding and diversity degree [16]. URLLC mainly reduces latency by shortening TTI and improves reliability by decreasing inter-user interference [4].

In URLLC, the requirement for user plane latency should be 0.5 ms for both uplink and downlink. In order to satisfy this, frame structure design of short TTI is introduced. There are two schemes for vertical service multiplexing single frame: one applies different TTI lengths for different vertical services; the other defines a short TTI, and all services use the set of this TTI. Since control channel is more reliable than data channel, new control channels need to be designed in URLLC. Frame structure design and vertical service multiplexing are introduced in the 3GPP RAN1 #86 meeting [17]. In order to coexist with other vertical services and reduce overhead, URLLC needs to further utilize HARQ mechanism.

NMA technology can realize multi-user signal transmission on same resource and obtain higher system spectrum efficiency. System robustness can be improved when access collision happens by introducing NOMA grant-free transmission schemes, whose performance is determined by variousness of NOMA scheme design, such as MCS choice, HARQ operation and so on.

When urgent data packet is sent by user, uplink URLLC transmission scheme start, as illustrated in Figure 4.5 [18]. Different from mMTC, URLLC cannot preset traffic, and the reserved UL resource is limited. When there is no reserved UL grant-free transmission resource for random access channel, uplink user will collide with other users, multiple access is then needed to handle UL resource conflict.

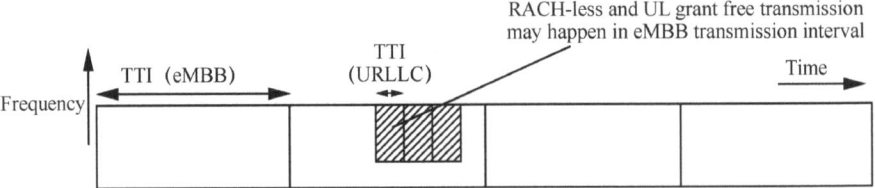

Figure 4.5: RACH-less and UL grant free in URLLC.

As illustrated is Figure 4.6, uplink grant-free transmission needs configured traffic between user and BS.

Reference [17] still introduces low latency data transmission in control channels and multicarrier spread spectrum-based design. The overall URLLC system design is to achieve reliability of 99.999% with 1 ms latency. Latency of each packet should include control channel, data transmission, processing, retransmission and queuing

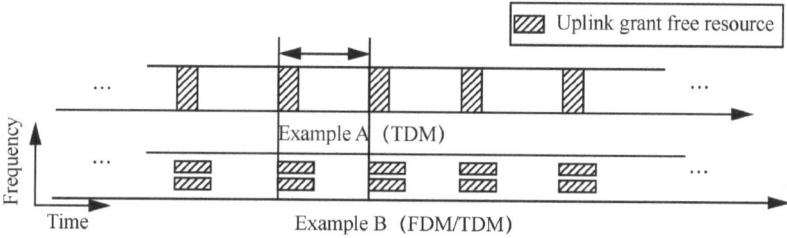

Figure 4.6: Uplink grant-free resource allocation in TDD.

latency. Hence, latency can be defined as the duration from the time when the packet arrives at MAC scheduler till the time when all bits of the packet are successfully decoded at MAC.

For downlink transmission, the latency should include following components:

$$L = L\_q + L\_Tx + L\_RxProc + (N-1) \times T\_HARQ\_RTT \tag{4.1}$$

where L_q is the queuing latency before a packet is scheduled; L_Tx is the transmission time; L_RxProc is the receiver processing time to decode the packet; N is the total number of transmission data; T_HARQ_RTT is the HARQ round trip time.

For uplink transmission, this latency should include following components:

$$L = L\_sr + L\_sr\_proc + L\_q + L\_grant + L\_grant\_proc + $$
$$L\_Tx + L\_RxProc + (N-1) \times T\_HARQ\_RTT \tag{4.2}$$

where L_sr is the scheduling request queuing and transmission time; L_sr_proc is the base station processing time for scheduling request; L_grant is the uplink grant transmission time; L_grant_proc is the uplink grant processing time. Figure 4.7 gives the analysis process of packet latency [19].

In order to actualize the uplink grant-free transmission, the corresponding uplink transmission format and procedures are the key. For the transmission format, a set of preambles, uplink control channel and data transmissions are shown in Figure 4.8 [18].
- Preamble: used for BS to detect uplink signal transmission.
- Uplink control channel: used to carry ID to identify UE, buffer status report, uplink control information and so on.
- Data transmission: to carry URLLC data and small packet.

Consider the case of NOMA application in URLLC. In order to support NOMA transmission, uplink resources for random access need to be statically preserved, and uplink grant-free transmission resources need to be semi-statically preserved. Uplink grant-free asynchronous transmission is performed within these preserved resources. Reference [20] describes the packet collision process in this situation in detail. It is

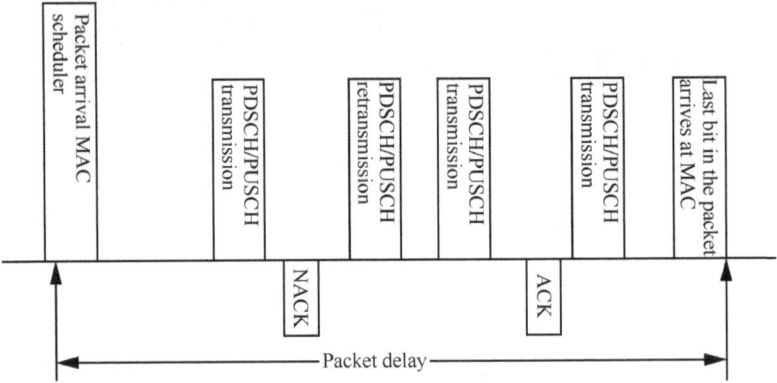

Figure 4.7: URLLC packet latency calculation.

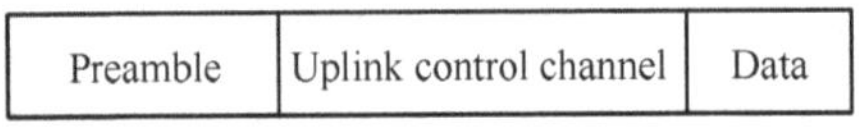

Preamble	Uplink control channel	Data

Figure 4.8: Example of transmission format for RACH-less and grant-free transmission.

hard for NOMA to handle such collisions. One of the possible ways is defining resource group with same time and frequency, and each resource group corresponds to certain coverage level. Each UE chooses UL resources based on separate measurement results, such as RSRP and path loss, as illustrated in Figure 4.9 [12].

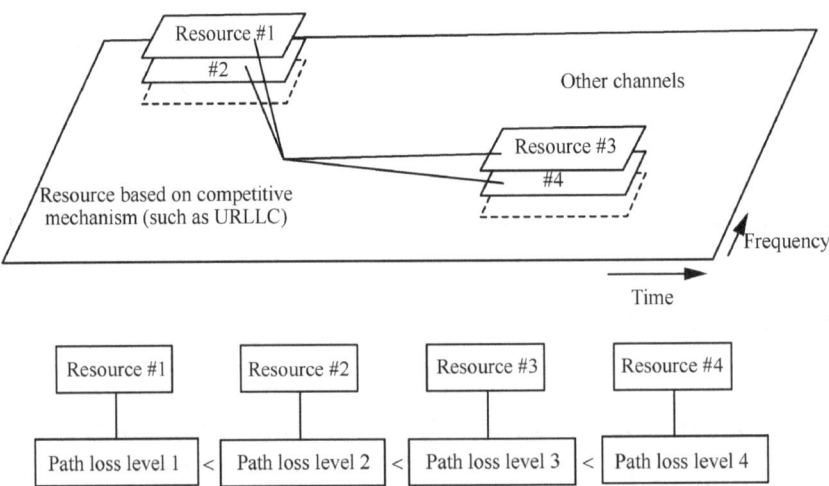

Figure 4.9: Uplink NOMA based on competitive transmission.

When urgent uplink data needs to be transmitted, UE starts uplink transmission without scheduling request immediately. For grant-free orthogonal multiple access, collision happens when users choose same time-frequency resources. Retransmissions caused by collisions will increase latency, and lead to transmission failure in case of serious collisions. Therefore, collision is against the requirements of URLLC. For grant-free NOMA, collisions may happen when users choose the same time-frequency resource or encoder/sequence/interleaver, but system robustness can still be guaranteed with advanced receivers [21].

4.2.3 Performance Evaluation of NMA in URLLC Scenario

Grant free non-orthogonal transmission improves reliability and reduces latency by decreasing collision probability. Reference [22] takes SCMA as an example and gives link-level simulation analysis to transmission reliability in URLLC. The simulation compares BLER performance of grant-free-based OFDMA and SCMA. Figure 4.10 is the BLER versus SNR link-level performance comparison between grant-free-based

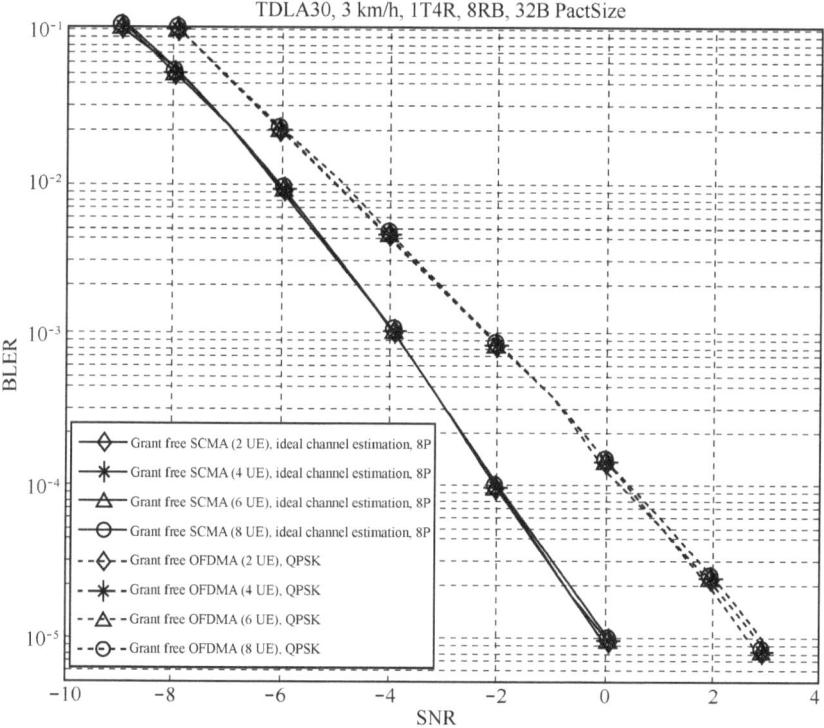

Figure 4.10: BLER versus SNR performance comparison between grant-free OFDMA and SCMA.

OFDMA and SCMA. From the figure, it can be seen that SCMA has obvious performance gain than OFDMA, which is because SCMA can obtain more diversity degree and some gain is brought by the encoding in SCMA codebook design. Table 4.2 is the SNR performance gain of grant-free SCMA compared with OFDMA when the reliability requirement is 99.999%. With grant free, SCMA can obtain a larger SNR performance gain than OFDMA, thus the BLER is lower with given SNR, which further indicates that grant-free-based SCMA is more suitable for URLLC scenario [23].

Table 4.2: SNR gain of grant-free SCMA over grant-free OFDMA.

2 UE	3.0 dB
4 UE	2.9 dB
6 UE	2.9 dB
8 UE	2.7 dB

4.3 Support for mMTC Scenario

Jie Zeng

4.3.1 mMTC Scenario Overview

mMTC is a typical application scenario in 5G, which is oriented to IoT and industry applications, supports massive connections and can truly realize the interconnection of all things [1]. Compared with traditional eMBB scenario and URLLC scenario, mMTC has its own characteristics, which mainly includes the following aspects.

(1) High Connection Density
mMTC is aimed at the IoT and various industry applications, and the terminal is MTC equipment. There are a variety of services included in IoT, such as intelligent meter reading, intelligent wearing, logistics tracking, health care, smart city and so on. Compared with traditional communications between people, terminals in IoT is much more. In addition, with the rise of the IoT and the emergence of new services, the number of IoT devices will grow explosively [24].

(2) Deep Coverage
For IoT system with fixed terminal positions, such as intelligent meter reading and so on, some MTC equipments may be installed in the basement and other places with poor channel conditions, thereby the channel between terminal and BS has

large fading. In order to cover such terminals, mMTC is required to support coverage enhancement to serve terminals with different positions.

(3) Low Power Consumption

Although different IoT applications have different requirements to terminal power consumption/battery lifespan, it is generally hoped to decrease the terminal power consumption to extend equipment standby time as long as possible. For example, for some environmental monitoring services, since some terminals cannot be charged, the battery lifespan determines the service life of MTC equipment. For intelligent wearable services, the power consumption requirement is not so strict, but it is still necessary to reduce the terminal power consumption as much as possible to improve user experience.

(4) Low Data Rate

Different application services in IoT have different characteristics. However, in general, the characteristics of mMTC services are low data rate, strong burst and large transmission interval. Such as intelligent meter reading, logistics tracking, health care and so on.

The specific Key Performance Indicator (KPI) proposed by 5G for the mMTC scenario is: the connection density is 1,000,000/ km^2, the maximum coupling loss supported by coverage is 164 dB, the battery lifespan of machine type equipment is up to 15 years, and the equipment implementation complexity is low.

In mMTC, competitive access-based users can be configured with one or multiple resource pools. Different users correspond to different resource pools. One resource pool includes a certain number of time-domain OFDM symbols and frequency subcarriers. In mMTC, one resource pool includes several frequency subcarriers and a large amount of time-domain TTI. Essentially, a resource pool includes one or multiple resource elements. mMTC does not need to define resource element type based on competitive transmission, but reuses resource type based on scheduling transmission. Different scenario cases correspond to different resource sizes. Figure 4.11 shows the resource pool structure in mMTC scenario [25], where a resource pool contains two resource elements.

In order to achieve the performance index of mMTC, NMA technology gives priority to grant-free schemes with low-cost and low power consumption. Grant-free scheme has the advantages of low signaling overhead, low latency and so on [26, 27].

Facing the urgent requirements and broad market prospect of IoT, there havebeen existed some IoT technologies. According to authorized or unauthorized spectrum application, these technologies can be divided into two categories. The first is the technologies applied to unauthorized spectrum, such as Long Range (LoRa), SigFox (one ultra-narrow band UNB technology) and so on; The second is the technologies applied to authorized spectrum, including 3GPP standardized Extended Coverage for GSM (EC-

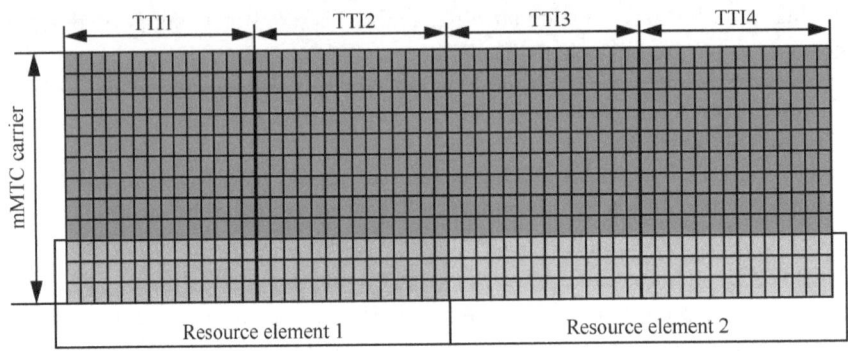

Figure 4.11: Resource pool structure in mMTC scenario.

GSM), Machine Type Communication (MTC) in Rel-12, enhanced Machine Type Communication (eMTC) in Rel-13 and Narrow Band Internet of Things (NB-IoT) in Rel-13.

The IoT technologies being standardized currently are Further enhanced MTC (FeMTC) and enhanced NodeB-IoT (eNodeB-IoT). EC-GSM is a further evolution based on GSM, which can support coverage enhancement. In Rel-12 MTC, a new UE type is introduced, which is Category 0 UE. Compared with traditional LTE UE, Category 0 UE can reduce UE complexity/cost by decreasing supported uplink and downlink rate and simplifying UE receiving radio frequency number. In Rel-13 eMTC, a new UE type, Category M1 UE, is introduced. Compared with Category 0 UE, Category M1 UE further reduces UE complexity/cost by reducing UE supported bandwidth. In addition, coverage enhancement is supported. Rel-12 MTC and Rel-13 eMTC have great inheritance to LTE, and only support working in LTE band. The new UE type introduced in Rel-13 NodeB-IoT is Category NodeB1, whose supported uplink and downlink radio frequency bandwidth and rate are further decreased. Besides working in LTE band, Category NodeB1 also supports working in LTE guard band or independent deployment.

Although 3GPP has carried out corresponding technical research and standardization for the application of IoT, there is still optimization space for requirements and KPI of 5G mMTC.

The uplink and downlink data transmission of current mMTC and NodeB-IoT is still based on BS scheduling, which means time-frequency resources used for uplink and downlink transmissions are allocated to UE semi-statically or dynamically. Semi-static allocation indicates Semi-Persistent Scheduling (SPS), in which BS pre-configures periodic time-frequency resources to UE, and further activates transmission on pre-configured resources by physical control channel; in dynamic allocation BS allocates time-frequency resources dynamically by physical control channel.

Transmission schemes based on BS scheduling is beneficial to network radio resource management, which is suitable for traditional MBB services. However, mMTC services have the characteristics of large connection number, small packet size and strong burst. The following problems will exist if transmission schemes based on scheduling is used.

1) SPS is applicable for periodic services and is not applicable for burst aperiodic MTC service;
2) Massive MTC connections will cause control channel limitation, thereby limit the scheduled user number;
3) For MTC small packet transmissions, the overhead of control channel is too large;
4) In transmission schemes based on scheduling, MTC terminals need to apply for resources from network first. Since most MTC services have strong suddenness and large transmission interval, this will cause further resource overhead.

Considering the MTC characteristics, such as small packet and strong burst, uplink grant-free-based transmission scheme is an important 5G research direction.

In the 3GPP RAN1 #86 meeting, following agreements are achieved. NR should support uplink grant-free scheme/mechanism at least in mMTC scenario, details need further study. Figure 4.12 describes a grant-free transmission scheme suitable for mMTC scenario.

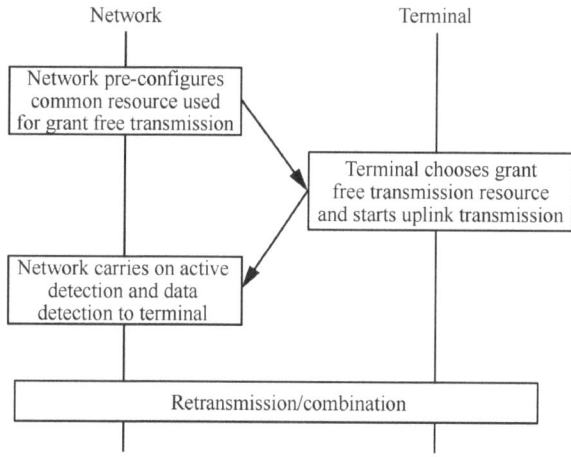

Figure 4.12: A candidate grant-free transmission scheme.

First, network configures common resources for grant-free transmissions, which are shared by UEs in the cell. When uplink data arrives, UE chooses resource from the common resources for grant-free transmissions configured by network and starts uplink transmission. There are two resource chosen methods for UE: one is that UE chooses grant-free transmission resource randomly; the other is to use pre-configured or pre-defined resource by the network. Since it is impossible for network to predict in advance whether there is UE and which UE starts uplink transmission on common resources used for grant-free transmission, activation detection and signal detection of UE are needed. Besides, since grant-free transmission resource is selected at the UE side, network needs to handle the collision case that multiple UEs choose same resource. By retransmission or HARQ, transmission reliability can be further improved.

Form TR 38.913 [1], it is known that there are massive mMTC equipment in urban environment. Massive transmission will cause the problem of connection efficiency. While limited resource will cause low connection efficiency, which means that more time-frequency resources are needed in mMTC scenario. Packet scheduling scheme divides users into several packets according to the same standard. If original user number is N, users can be grouped according to specific indicators, which may be channel quality, transmission time and SINR. As illustrated in Figure 4.13, users between two dashed lines means they have same transmission conditions. Figure 4.13(a) is the case without packet scheduling, while Figure 4.13(b) is the case with packet scheduling. Besides, users in the same group occupy same time-frequency resource by SPS, and they are scheduled on one or several resource blocks to reduce congestion and collision. BS does not need to send uplink scheduling grant to UE. Access user number reduces from N to K. Access efficiency is the most significant if $K = \log N$.

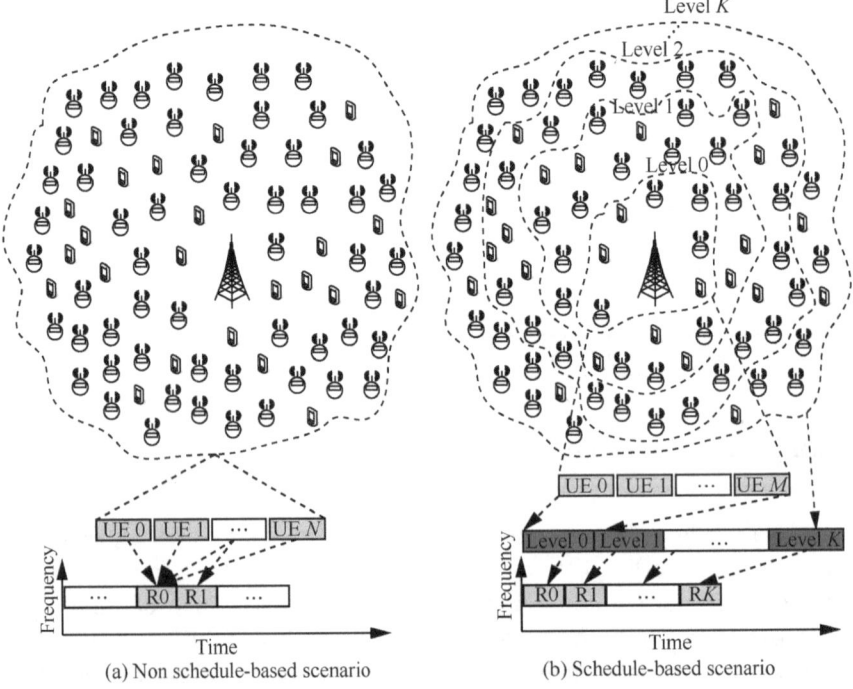

(a) Non schedule-based scenario   (b) Schedule-based scenario

Figure 4.13: Group scheduling.

Users in each layer apply grant-free schemes based on multiple access [28]. Possible access process is as follows.

1) UE → eNodeB
UE sends information to eNodeB through uplink control channel, such as channel information, transmission time, power and so on.

2) eNodeB → UE
eNodeB divides UEs into several groups according to certain criteria, and sends reference information to certain UE ID and TTI. Then eNodeB allocates time-frequency resource with same level.

3) UE → eNodeB
First UE decodes reference signal by specific mMTC control channel, then UEs within same group in different TTI transmit data in specific slot. This could avoid base station scheduling and reduce control channel overhead.

Uplink and downlink data transmissions in current MTC and NodeB-IoT technology are based on orthogonal multiple access, in which different users transmit on orthogonal time-frequency resources.

Since orthogonal multiple access technology cannot support multiple UE sharing same time-frequency resources, supported user number on certain time-frequency resources is limited. Furthermore, if combination with uplink grant-free transmission is considered and only orthogonal multiple access is supported, as the uplink transmission time at the UE is random, in order to guarantee same time-frequency resource is not chosen by multiple UEs, network needs to pre-configure dedicated time-frequency resources for UE. MTC services have strong burst and long transmission interval, the way to pre-configure dedicated resources will waste wireless resources greatly and cannot satisfy the KPI requirement of 5G connection density.

In order to overcome the above limitations of orthogonal multiple access in mMTC, 5G will support NOMA. By superposition transmission of multi-user information, system connection ability can be doubled. In addition, the overhead and terminal power consumption can be effectively reduced by grant-free transmission.

4.3.2 Applications of NMA in mMTC Scenario

The most important KPI in mMTC scenario is massive connection support [1]. In 3GPP RAN1 meetings, several NMA schemes have been proposed to support large number of equipment, including MUSA, RSMA, SCMA, PDMA, IDMA, IGMA and so on [29].

Grant-free scheme/mechanism is the best choice to satisfy mMTC requirements such as low cost, low power consumption and so on. NOMA technologies based on autonomy/grant free/competition scheme/ mechanism need to meet the following two points.

1) UE transmission does not need dynamic and explicit scheduling grant of the network. For example, SPS, RACH-free and competition-based access.
2) Multiple users can share same time-frequency resources.

In SPS-based access, uplink transmission users allocate PRB resources. The characteristics of SPS are control signaling reduction and resource allocation support for small packet such as VoIP. However, in mMTC scenario, most communications are non-continuous, which means there are few packets with long term activities. Most communication data are low rate small packet. In order to support massive connections, each UE must release connection. Once UE completes data transmission, the system goes into sleep state.

The real grant-free transmission should be RACH-free, which means that UE packet transmission is allowed at any time. In NR multiple access candidate schemes, non-orthogonality is mainly based on code domain spreading or cross diversity technology. Reference [30] proposes the grant-free MUSA, as illustrated in Figure 4.14. UE chooses a coding sequence randomly form pre-configured resource pool. Once small packet transmission is successful, UE turns to sleep state and does not need RACH and closed loop power control. At the base station, dynamic scheduling grant is not needed, base station does not know active user number, spreading code of each user or channel information [31].

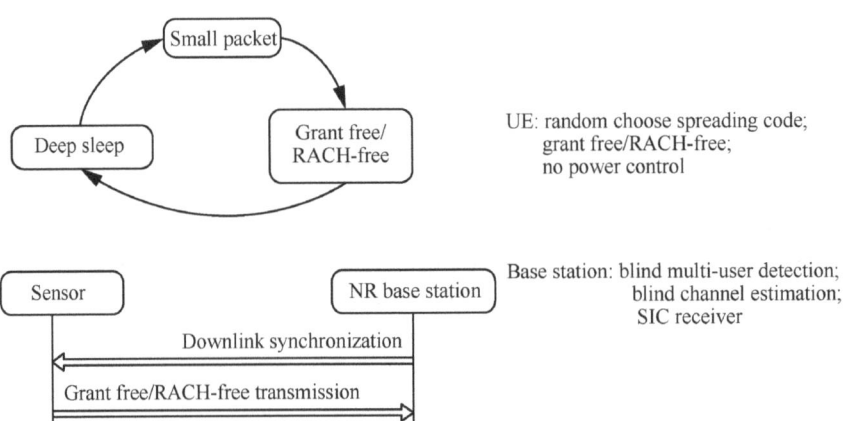

Figure 4.14: Grant-free MUSA in mMTC scenario.

In 5G mMTC scenario, grant-free MUSA is very desirable. In terms of power saving, UE can send data packet at any time, and return to deep sleep state immediately when there is no data packet transmission; low-cost design at the UE side, BS only needs to consider the receiving complexity; no scheduling, signaling overhead and transmission latency are reduced.

In mMTC, uplink multiple access schemes are mainly used in low SNR and low spectrum efficiency systems. In terms of spectrum efficiency, users with high spectrum efficiency do not need mMTC. System design is important. By developing typical mMTC traffic equipment, maximum system resource utilization efficiency is achieved. A good NOMA design scheme can improve the transmission throughput gain of non-orthogonal users. In dealing with a large number of connection requirements, the key of NOMA design is analyzing form channel estimation and data group decoding, and reload physical resources.

Large area coverage and deep service are the requirements of mMTC scenario application cases, which reflect the robustness of NOMA in channel estimation accuracy, time offset and frequency offset. If low quality components are used in mMTC equipment, system performance can be affected. For example, low-cost oscillator may cause extra frequency offset and poor UE tracking ability [26].

In NR systems, multiple access schemes must satisfy changing requirements in different application scenarios. In mMTC scenario, it can control signaling overhead, mMTC user power storage and connection density. Specifically, data packet in mMTC is small, thereby compared with other scenarios, the key problem is signaling overhead control.

For the above purposes, mMTC users attempt to use contention-based UL transmission. UEs share a group of common NOMA resources and choose one resource randomly for UL transmission. Then, eNodeB reduces inter-user interference by advanced receivers [27, 32].

4.3.3 Performance Evaluation of NMA in mMTC Scenario

In mMTC scenario, in order to support large MTC traffic, competition-based multiple access needs to be used. Competition-based multiple access means that scheduling is not needed for any initial data packet transmission or following retransmission.

Consider 5G deployment scenarios, operators have 10 MHz bandwidth to support broadband services and mMTC services. It is assumed that mMTC load is statistically equivalent to each cell (such as λ Poisson arrival/s). Part of the spectrum is used for mMTC service, which is the same area in each cell, while others is left for wideband network traffic. Therefore, as the traffic load of mMTC increases, mMTC needs more spectrum.

Reference [33] gives analysis to competition-based uplink NOMA technologies in mMTC, and evaluates the throughput metrics in the mMTC scenario. Simulation parameters are listed in Table 4.3.

Table 4.3: Simulation parameters.

Parameter	Value
Carrier frequency	2 GHz
Bandwidth of each PRB	180 kHz
TTI	1 ms
Topology	19 sites, 3cells per site, ISD = 1,732 m
Channel coding	Turbo coding
Channel model	TU
UE speed	3 km/h
HARQ	yes
MCS	QPSK, 1/2
UE maximum power	mMTC: 13 dBm, broadband: 23 dBm
Open loop power control	mMTC: $\alpha = 1$, optimal $P0$; broadband: $\alpha = 1$, optimal $P0$

With this model, Figure 4.15 shows the throughput changes of broadband and mMTC in different mMTC load bandwidths when the inter station spacing is 1,732 m. Results indicate that the broadband throughput is high, and high mMTC load has a significant impact on system throughput, which requires higher spectrum efficiency.

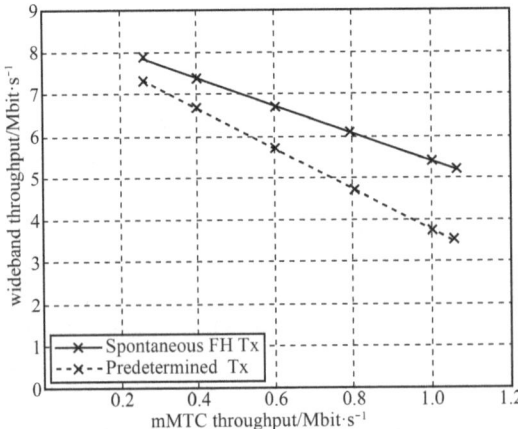

Figure 4.15: Performance comparison between frequency hopping competitive access and pre-defined access.

In IMT-2020 multiple access technical report, system-level evaluation of grant-free-based MUSA in mMTC scenario is proposed. Simulation parameters are listed in Table 4.4 [2].

Table 4.4: System-level simulation parameters in mMTC scenario.

Parameter	Value
Layout	Hexagonal cell, 19 cells with 3 sectors per cell, ISD = 500 m
Channel scenario	ITU UMa
Carrier frequency	2 GHz
Occupied bandwidth	4 PRB (720 kHz)
UE number in each cell	Small packet, inter-arrival time is a Poisson distribution
Proportion of indoor users	0%
UE speed	3 km/h
UE power control	Open loop power control, $\alpha = 1$, $P0 = -95$ dBm, maximum transmit power 23 dBm
Antenna number	1T2R
Antenna configuration	UE: vertical polarization. eNodeB: $\pm 45°$ cross-polarization
Antenna mode	3D, same as TR 36.814
Antenna height	eNodeB: 25 m. UE: 1.5 m
Antenna gain and link loss	17 dBi
UE antenna gain	0 dBi
eNodeB antenna tilt	ITU Uma: 12°
Minimum horizontal distance between UE and eNodeB	ITU Uma: 35 m
Cell selection criteria	RSRP
Handover margin	0 dB
Network synchronization	Synchronous
Channel estimation	Ideal
MCS mechanism	Fixed MCS, grant free
Maximum HARQ number	No
Traffic model	FTP1 (refer to TR 36.814), packet size 144 bit
Performance measurement	Supported load with 1% packet loss rate

Reference [2] describes specific assessment method. In the simulation, it is assumed that UE adopts grant-free transmission schemes. When different UEs occupy same RE or pattern, collision happens. For OFDMA, collided signals are regarded as interference with each other. For MUSA, BS carries on collision detection with advanced receivers. Simulation results are illustrated in Figure 4.16. Non-orthogonal MUSA has significant performance improvement in packet loss rate compared with OFDMA.

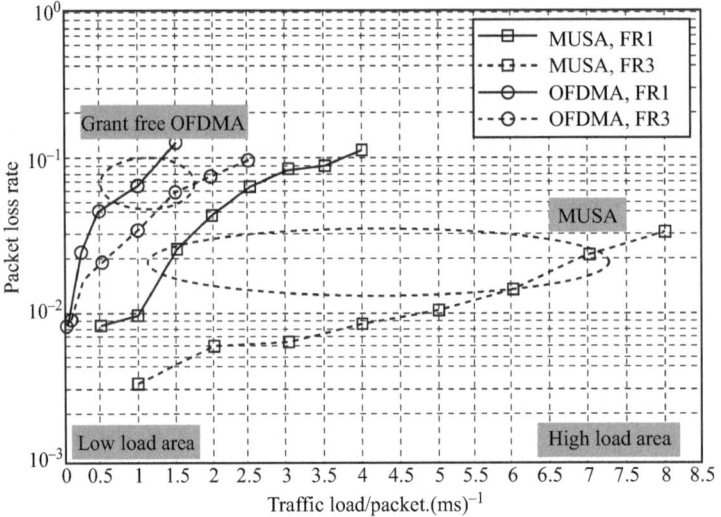

Figure 4.16: System-level simulation results comparison.

References

[1]   3GPP TR 38.913. Study on scenarios and requirements for next generation access technologies[S]. 2016.
[2]   R1-162226. Discussion on multiple access for new radio interface[R]. ZTE, 3GPP TSG RAN WG1 Meeting #84bis, Busan, Korea, 2016.
[3]   R1-167393. Discussion on multiple access for eMBB[R]. NTT DOCOMO, INC, 3GPP TSG RAN WG1 Meeting #86, Gothenburg, Sweden, 2016.
[4]   R1-162153. Overview of non-orthogonal multiple access for 5G[R]. Huawei, HsiSilicon, 3GPP TSG RAN WG1 Meeting #84bis, Busan, Korea, 2016.
[5]   R1-164036. Multiple access for UL small packet transmission[R]. Huawei, HiSilicon, 3GPP TSG RAN WG1 Meeting #85, Nanjing, China, 2016.
[6]   R1-164246. Discussion on scenarios and use cases for multiple access[R]. CATT, 3GPP TSG RAN WG1 Meeting #85, Nanjing, China, 2016.
[7]   R1-162870. On unified framework for multiple access schemes[R]. CMCC, 3GPP TSG RAN WG1 Meeting #84bis, Busan, Korea, 2016.

[8]   R1-162155. Sparse code multiple access (SCMA) for 5G radio transmission[R]. Huawei, HiSilicon, 3GPP TSG RAN WG1 Meeting #84bis, Busan, Korea, 2016.

[9]   R1-1610076. Discussion on multiple access for eMBB[R]. NTT DOCOMO, INC., 3GPP TSG RAN WG1 Meeting #86bis, Lisbon, Portugal, 2016.

[10] R1-164178. Uplink non-orthogonal multiple access for nr access technology[R]. Intel Corporation, 3GPP TSG RAN WG1 Meeting #85, Nanjing, P.R.O.C, 2016.

[11] R1-167336. SLS result for SCMA in DL eMBB scenario[R]. Huawei, HiSilicon, 3GPP TSG RAN WG1 Meeting #86, Gothenburg, Sweden, 2016.

[12] R1-165175. Initial views and evaluation results on non-orthogonal multiple access for NR[R]. NTT DOCOMO, INC., 3GPP TSG RAN WG1 Meeting #85, Nanjing, China, 2016.

[13] R1-162305. Multiple access for 5G new radio interface[R]. CATT, 3GPP TSG RAN WG1 Meeting #84bis, Busan, Korea, 2016.

[14] ITU-R M.2083: IMT Vision. Framework and overall objectives of the future development of IMT for 2020 and beyond[S]. 2015.

[15] R1-1609300. Discussion on multiple access schemes for URLLC[R]. CMCC, 3GPP TSG RAN WG1 Meeting #86bis, Lisbo, Portugal, 2016.

[16] R1-162189. Ultra-reliability with low-latency support in 5G new radio interface[R]. Samsung, 3GPP TSG RAN WG1 Meeting #84bis, Busan, Korea, 2016.

[17] R1-166493. Ultra-low latency scheduling-based UL access[R]. Idaho National Laboratory, 3GPP TSG RAN WG1 Meeting #86, Gothenburg, Sweden, 2016.

[18] R1-165174. Uplink multiple access schemes for NR[R]. NTT DOCOMO, INC., 3GPP TSG RAN WG1 Meeting #85, Nanjing, China, 2016.

[19] R1-166398. URLLC system level simulation assumptions[R]. Qualcomm Incorporated, 3GPP TSG RAN WG1 Meeting #86, Gothenburg, Sweden, 2016.

[20] R1-167392. Discussion on multiple access for UL mMTC[R]. NTT DOCOMO, INC., 3GPP TSG RAN WG1 Meeting #86, Gothenburg, Sweden, 2016.

[21] R1-166466. Usage scenarios of non-orthogonal multiple access[R]. CATT, 3GPP TSG RAN WG1 Meeting #86, Gothenburg, Sweden, 2016.

[22] R1-1608869. Grant-free non-orthogonal MA for uplink URLLC[R]. Huawei, HiSilicon, 3GPP TSG RAN WG1 Meeting #86bis, Lisbon, Portugal, 2016.

[23] R1-166359. Resource spread multiple access[R]. Qualcomm Incorporated, 3GPP TSG RAN WG1 Meeting #86, Gothenburg, Sweden, 2016.

[24] R1-162224. Evaluation assumptions for NR in RAN1[R]. ZTE, 3GPP TSG RAN WG1 Meeting #84bis, Busan, South Korea, 2016.

[25] R1-167254. Channel structure for contention based access[R]. Nokia, Alcatel-Lucent Shanghai Bell, 3GPP TSG RAN WG1 Meeting #86, Gothenburg, Sweden, 2016.

[26] R1-166551. Views on UL multiple access for NR[R]. Intel Corporation, 3GPP TSG RAN WG1 Meeting #86, Gothenburg, Sweden, 2016.

[27] R1-166873. Discussion on categorization of MA schemes and target scenarios[R]. LG Electronics, 3GPP TSG RAN WG1 Meeting #86, Gothenburg, Sweden, 2016.

[28] R1-167742. Discussion on multiple access for mMTC[R]. Institute for Information Industry (III), 3GPP TSG RAN WG1 Meeting #86, Gothenburg, Sweden, 2016.

[29] R1-166403. Grant-free multiple access schemes for mMTC[R]. ZTE, ZTE Microelectronics, 3GPP TSG RAN WG1 Meeting #86, Gothenburg, Sweden, 2016.

[30] R1-164270. Receiver implementation for MUSA[R]. ZTE, 3GPP TSG RAN WG1 Meeting #85, Nanjing, China, 2016.

[31] R1-166405. Discussion on grant-free concept for UL mMTC[R]. ZTE, ZTE Microelectronics, 3GPP TSG RAN WG1 Meeting #86, Gothenburg, Sweden, 2016.

[32] R1-166875. Considerations on the receiver types and NoMA schemes[R]. LG Electronics, 3GPP TSG RAN WG1 Meeting #86, Gothenburg, Sweden, 2016.

[33] R1-167248. Contention-based non-orthogonal multiple access with frequency hopping for the mMTC uplink[R]. Nokia, Alcatel- Lucent Shanghai Bell, 3GPP TSG RAN WG1 Meeting #86, Gothenberg, Sweden, 2016.

Index

https://doi.org/10.1515/9783110666366-005